CAMBRIDGE LATIN AMERICAN STUDIES

GENERAL EDITOR
MALCOLM DEAS

ADVISORY COMMITTEE
WERNER BAER   MARVIN BERNSTEIN
AL STEPAN   BRYAN ROBERTS

43

JUAN VICENTE GOMEZ
AND THE OIL COMPANIES
IN VENEZUELA, 1908–1935

JUAN VICENTE GOMEZ AND THE OIL COMPANIES IN VENEZUELA, 1908–1935

B.S. McBETH

CAMBRIDGE UNIVERSITY PRESS

Cambridge
London   New York   New Rochelle
Melbourne   Sydney

338.27282

M11j

Published by the Press Syndicate of the University of Cambridge
The Pitt Building, Trumpington Street, Cambridge CB2 1RP
32 East 57th Street, New York, NY 10022, USA
296 Beaconsfield Parade, Middle Park, Melbourne 3206, Australia

First published 1983

Printed in Great Britain at
the University Press, Cambridge

Library of Congress catalogue card number: 82-14667

British Library Cataloguing in Publication Data
McBeth, B.S.
Juan Vicente Gómez and the oil companies in
Venezuela, 1908–1935.—(Cambridge Latin American
studies; 43)
1. Petroleum industry and trade—Venezuela—
History   2. Industry and state—Venezuela
—History
I. Title
338.2'7282'0987   HD9574.V42

ISBN 0 521 24717 9

dyd

RB

Contents

Tables

Map

Preface

This book is not a history of Gómez's dictatorship but examines specifically the relationship between the Gómez regime and the oil industry which developed during his government. I am most grateful to my parents, to whom the book is dedicated, for their support in the preparation of this work. As always, my wife María Cristina, was very helpful at every stage in the preparation of this book. I would also like to thank Malcolm Deas for his detailed and meticulous comments on the manuscript and for his general encouragement. I am also indebted to Senator Dr Ramón J. Velásquez for his hospitality and counsel in Caracas, and for gaining me access to the Archivo Histórico de Miraflores and to the Archivo dependiente de la División de Conservación de la Oficina Técnica de Hidrocarburos (Ministerio de Energía y Minas) for my studies. I wish also to thank the following people in Caracas: Dr Gumersindo Torres Ellul, Professor Manuel Pérez Vila, Dr Pedro Grases, Dr César González, Dr Nikita Harwich, and last but not least, Dr and Mrs William M. Sullivan for their kind hospitality and help. A special thanks is also due to the staff of the Centro de Estudios Latinoamericanos 'Rómulo Gallegos' who bore my many requests with great patience and tact.

My stay in Caracas was financed by a Foreign Area Fellowship granted by the Social Science Research Council (U.S.) and the American Council of Learned Societies, and by grants from the Centro de Estudios Latinoamericanos 'Rómulo Gallegos', Fundación John Boulton, and St Antony's College, Oxford.

The conclusion, opinions, and other statements in this work are those of the author, and not necessarily those of the above mentioned institutions. All quotations from the original Spanish have been translated by the author. For any errors of omission or commission I alone am accountable.

ix

Abbreviations

AAAC	Archivo del General Antonio Aranguren. Carpeta
ADCOTHMEM	Archivo dependiente de la División de Conservación de la Oficina Técnica de Hidrocarburos, Ministerio de Energía y Minas
AMF	Archivo del Ministerio de Fomento
AMFHC	Archivo del Ministerio de Fomento. Hidrocarburos, Correspondencia, uncatalogued
AGTC	Archivo particular del Dr Gumersindo Torres. Correspondencia
AGTCOP	Archivo particular del Dr Gumersindo Torres. Copiador
AGTAPM	Archivo particular del Dr Gumersindo Torres. Artículos, Proyectos y Memorándums
AGTU	Archivo particular del Dr Gumersindo Torres. Unclassified
AHMCOP	Archivo Histórico de Miraflores. Copiador
AHMSGPRCP	Archivo Histórico de Miraflores, Secretaría General de la Presidencia. Correspondencia Presidencial
AHMSGPRCS	Archivo Histórico de Miraflores, Secretaría General de la Presidencia. Correspondencia del Secretario General
AHMSGPRCPU	Archivo Histórico de Miraflores, Secretaría General de la Presidencia. Correspondencia Presidencial, unclassified
AHMV	Archivo Histórico de Miraflores. Varios
APACT	Archivo del Dr Pedro A. Capiola Torres

BAHM	*Boletín del Archivo Histórico de Miraflores*
BCCC	*Boletín de la Cámara de Comercio de Caracas*
Bs.	bolívars
CAB	Cabinet Papers, Public Record Office
CANIP	C.A. Nacional de Industrias Petroleras
CDC	Colon Development Co. Ltd
CPC	Caribbean Petroleum Company Ltd
CVP	Compañía Venezolana de Petróleo S.A.
DDCD	Diario de Debates de la Cámara de Diputados
DDCS	Diario de Debates de la Cámara del Senado
DS	Department of State
Exxon	Standard Oil Company (New Jersey)
FO	Foreign Office, Public Record Office
Gulf Oil	Venezuelan Gulf Oil Company
Lago	Lago Petroleum Company
MinFo	Ministerio de Fomento
MinHa	Ministerio de Hacienda
MinRelExt	Ministerio de Relaciones Exteriores
MinRelInt	Ministerio de Relaciones Interiores
PA	Pearson Archive
PP	Parliamentary Papers
RLDV	Recopilación de Leyes y Decretos de Venezuela
Shell Group	Royal Dutch-Shell Group
SOC(CAL)	Standard Oil Company (California)
SOC(IND)	Standard Oil Company (Indiana)
SOC(Venezuela)	Standard Oil Company of Venezuela
VJRCA	Venezuela, Jurado de Responsabilidad Civil y Administrativa
VOC	Venezuelan Oil Concessions Ltd

Introduction

The oil companies were attracted to Venezuela because of the expectation of large oil deposits, relative political stability, and the good terms offered for the exploitation of the country's oil resources. Venezuela's geographical position *vis à vis* the major oil markets placed her at a freight advantage over both her Mexican and Middle East competitors, and even over some American oilfields. Maracaibo is 113 miles nearer New York, and 863 miles nearer Southampton, than the Mexican port Tampico, and is only 644 miles from the Panama Canal.[1] This meant that Venezuela was 'favourably situated to export to Europe, to the United States and to the Pacific coast of America, and the Far East via the Panama Canal'.[2] Furthermore, her logistic production problems were far fewer than those of her Middle East competitors, because the oilfields were conveniently situated near sea transport, and because the offshore Dutch islands of Curacao and Aruba provided the companies with a safe haven in which to build their large refineries. Unlike the Middle East and Iran, Venezuela devised a concessionary system whereby most oil companies, regardless of nationality, could operate. In addition, production costs were much lower than in the U.S. and the number of barrels produced per active well much higher. Moreover, 80.5 per cent of the total number of wells drilled in Venezuela up to 1935 were productive, thus demonstrating much lower exploration and drilling costs in Venezuela than in the U.S.[3] As a result, the average cost of producing a barrel of Venezuelan oil was $0.65, compared to $1.98 in the U.S.[4]

The development of the Venezuelan oilfields offered the companies higher profits, cheap oil, and the flexibility of being able to adjust output and distribution to changing world

1

2   *Gómez and the oil companies in Venezuela, 1908–1935*

Table 1   *Number of barrels produced per active well in Venezuela and the U.S. in 1928*

Region	Crude oil production (in million barrels)	No. of active wells	Barrels per active well (in thousand barrels)
Texas	255.4	910	280.7
Oklahoma	247.5	715	346.2
California	232.0	511	454.0
VENEZUELA	106.5	85	1,252.9
Kansas	38.2	225	169.8
Arkansas	32.1	62	517.7

Source: adapted from DS 831.6363/149, E.B. Hopkins, 'Some geological and economic notes on the Venezuelan oil developments', Paper given at the American Association of Petroleum Geologists, Fort Worth Meeting, March 1929

markets. The large vertically integrated oil company was interested in stabilizing the domestic U.S. market at a relatively high price for crude oil, thus allowing the company to supply foreign markets with cheap Venezuelan oil but to charge the U.S. high prices through the 'Gulf +' pricing structure then in use. This system effectively barred the cheap Venezuelan crude from competing with domestic American oil, and therefore prevented it from disturbing the American domestic market. But as oil production was controlled by the major oil companies and Venezuelan interests did not participate in the international oil markets, this effect was unimportant. The system did, however, have two overriding advantages for the continual growth of the young Venezuelan oil industry, which linked the interests of the Venezuelan government with those of the large vertically integrated oil companies. First, the country was assured of continuously rising oil revenues because the large oil operators would use Venezuelan oil as a lever to control the U.S. domestic oil industry and as a supply for European and other world oil markets; and, second, high U.S. oil prices meant that government royalty payments (expressed as a percentage of the value of oil) would be greater. The companies operating in

Venezuela also achieved considerable success, since, with the introduction of prorationing measures in the early 1930s, they had, within the United States, the structure to achieve and maintain high oil prices, while Venezuelan oil could be used to control any new major fields which might develop, and at the same time the large European markets could be supplied with cheap Venezuelan oil at high American oil prices.

The development of the oil industry in Venezuela took place under the rule of General Juan Vicente Gómez, who came to power in a bloodless coup on 19 December 1908, and died in his sleep on 17 December 1935. During his 27 years as absolute ruler, from being a predominantly agricultural country, Venezuela became the second largest oil producer in the world. Before this Venezuela was little known to the outside world, but by Gómez's death she had become of vital strategic importance to the British Empire, and in addition an important supplier of oil to America's Atlantic seaboard.

The political and economic problems which Gómez faced at the beginning of his rule were considerable. His economic plans were very ambitious, given the backwardness of the country's economic infrastructure and Venezuela's reputation in the major international money markets. In order to secure peace and stability during the ensuing years Gómez would have to maintain a delicate neutrality between the various political factions which were already claiming him as their true leader. At the same time he would have to stimulate the economic development of the country to a degree which had not been achieved previously. He was well aware of the economic constraints operating in the country and of the adverse influence which the German trading houses could exert on the country's economy. It was therefore necessary to stimulate the development of an independent source of revenue free from traditional political considerations. Consequently, from the outset of his rule Gómez encouraged the establishment of a healthy and thriving mining industry. There was nothing new in this idea – past rulers had also pinned their hopes on large mining revenues. What was novel in Gómez's case was that he achieved this independent source of income with the advent of the oil industry in the 1920s.

The relationship between oil companies and governments is

one of continuous adjustments as a result of changes in the international oil markets, the local economy, and local politics, with the government being in a fundamentally weaker position than the oil companies. This book examines this relationship, looking at the government's initial reaction to the new oil industry, the legal framework created to control and supervise it, the socio-economic effects, both locally and nationally, and the degree of control exercised by the government over the exploitation of its national resource.

1
The dawning of an era

When Gómez seized power in 1908 his underlying objectives were to bring peace and work to Venezuela. By engineering his political supremacy he managed to secure peace, but the latter objective would be ensured only by laying the foundation for a stronger and more modern economy. One of the first measures he took was the improvement of the health and sanitation conditions of the country. In addition, education was stimulated and reorganized under the 1912 Código de Instrucción. From the fiscal chaos under Castro the National Treasury was also restructured, a process which culminated in 1918 with the enactment of the first Ley Orgánica de la Hacienda Nacional. At the same time Gómez strengthened his own position by establishing in 1910 the Inspectoría General del Ejército, 'so that the army would be under constant and operational supervision',[1] and also the Military and Naval Academies. However, the greatest change which occurred in these initial years resulted from the stimulus given to road construction. The country's transport system, as Gómez had witnessed during his past military campaigns, was in an appalling state; roads were almost non-existent, and the easiest and fastest form of travel was by sea, or by the small rail network where it existed. Consequently, in order to knit the country together, and thus bring greater unity and prosperity, on 24 June 1910 Gómez initiated a massive programme of road building and repair work, with half the Ministry of Public Works' annual budget devoted to the building of new roads.[2] The difference which this made to the country was soon apparent. Táchira, so long isolated from the rest of the country, was linked to central Venezuela in 1915 by the completion of part of the Carretera Central de Táchira.

In his first Annual Message Gómez called on Congress to enact laws that would expand credit facilities for industry and foster the development of the country's agriculture (by the creation of an agricultural development bank) and mining resources more fully.[3] The inadequate credit facilities, and the innate reluctance on the part of the commercial elite to take risks, meant that Gómez's 'economic-social'[4] plan depended heavily on foreign capital to develop the country's natural resources. Owing to the special conditions of the country, foreign capital would serve as a guarantee to native capital in any large enterprise, and it would also serve to further consolidate Gómez's rule. In order to increase the credit-worthiness of the country still further, in 1912 Gómez paid off the outstanding debt of Bs.4 million which remained of the Washington Protocols. The tremendous increase in trade which the country experienced in these early years helped him to achieve this end, and this was accompanied *pari passu* by growth in foreign investment.

THE MINING INDUSTRY

From the very beginning of Gómez's rule there was a keen interest in developing the mineral wealth of the country, for it was felt that this would spearhead its general development, and provide an independent source of income. In order to make Venezuela's natural resources better known to the world it was decided that an Exploitation Commission, composed of experts from various European countries and the U.S., should be formed to inform 'European industry of the possibilities which exist to exploit our minerals'.[5] In addition, Venezuelan ministers at the various legations were instructed to promote the country's natural resources as a good investment for foreign capital. Local capital was also anxious to develop the mineral wealth of the country. For instance, soon after the December 1908 coup, Dr Pedro Miguel Queremal and General R. Cayama Martínez travelled to Caracas to confer with Gómez over the possibility of exploiting the coal deposits of Falcón State.[6]

However, before foreign capital could be attracted to the country in large quantities, there was an urgent need to reform the existing 1906 Mining Code which discouraged foreign capital from entering the country. Hence on 11 August 1909 a new Mining Code was enacted, which was more liberal in its terms, and which ratified the 'dominial' system started under the 1904 Code, and also increased government supervision over mining activities. Mining taxes were reduced from 3 to 1 per cent of gross production, and the annual surface tax was reduced from Bs.2 to Bs.0.50 per hectare. The new law established the principle of granting exoneration of customs duties on mining equipment automatically, and also revalidated all existing titles, even those which had recently expired. A tribunal in which both parties were represented would now decide when a title was cancelled, in place of the old system which did not provide the owner with the right of defence or appeal. The government gave the mining industry a further stimulus when, on 30 September 1909, it wrote off all outstanding mining taxes. The law was greeted with approval. Through it industrial expansion would be encouraged, which in turn would curb the country's large unemployment problem, but above all, it was welcomed because of its inducements to foreign capital.

The 1909 Mining Law did have one flaw: it subjected all concessions granted under previous laws to the payment of new taxes. So on 23 February 1910 the Federal Court of Cassation declared that this clause was null and void and upheld the view that no other law but that under which the mining title had been granted could be applied to any particular mine. Moreover, it was also felt that the law did not provide sufficient financial inducements to the investor. Gómez was therefore forced to request Congress to revise the law once again.

The bill which Congress debated allowed for very moderate taxation, which Lisandro Alvarado, President of the Senate, considered desirable since the benefits to the country from the mining industry would not be in revenue but in the 'increase in the number of firms established'[7] as a result of the mining industry. In June 1910 Congress approved the new law, which contained all these provisions but which also increased taxation from 1 to 3 per cent of gross production. According to Alvarado, the law was a just reflection of Gómez's economic aspirations for the country because it would open a new period of 'investment, immigration, and roads'.[8]

It was important that proper legislation should be enacted to guarantee the development of the industry, and Gómez and his government pinned their hopes increasingly on the additional revenue which the mining industry would bring. Although the country was by no means a 'concession hunters' paradise',[9] Sir Vincent Corbett, the British Minister, observed that the number of concessions awarded had increased considerably since Gómez's rise to power.[10] This increase was mainly due to his policies to stimulate mining, as well as to economic prosperity in general, but also because it was much easier now to obtain concessions. As we have seen, the 1909 Mining Law gave the Executive greater powers to negotiate mining contracts in general. This was an extension of the 'dominial' principle established in the 1904 Mining Code, which placed the exploitation of asphalt, oil, and similar substances under special government contracts. The difference here was that under previous laws the first denouncer of a mine would acquire the title automatically; now the Executive, without the approval of Congress, would decide the terms and conditions under which the deposits were to be exploited. The government could therefore enter directly into oil exploitation or lease this right to a third party. The 1909 and 1910 Mining Laws ratified this principle.

Without checks and balances the system was open to corruption since the Executive could now deal in the country's resources for personal gain. Under the 1909 Constitution, however, the Gómez Administration had to consult with the Council of State, and gain its approval for all negotiations and contracts entered into by the Executive. But, as we shall see, this did not prevent the Executive from entering into corrupt transactions. For example, Ascanio Negrette, in order to get his title to El Mene asphalt mine in Zulia confirmed, agreed to pay Gómez 50 per cent of the net profits made from selling the mine to foreign investors.[11]

Despite the 1909 and 1910 changes, the law still contained a serious defect, which was that it awarded the surface owner 33.3 per cent of the gross production of any mining activity which took place on the owner's land. Consequently, the Attorney-General presented a petition at the Federal Court of Cassation to have the relevant articles declared invalid, which was done on 27 June 1912. Control of the mining sector was further strengthened when on 1 July 1913 the post of Inspector-General of Mines was established, and a new department within the Ministry of Development, the Sala Técnica de Minas was created, incorporating the offices of the Technical Inspector of Mines, the Inspector-General of Mines and the Director of the National Laboratory. Finally, a Ministerial Resolution of 19 August 1913 called for more details to be given to the Servico de Guardaminas by the mining companies about the working conditions and geological formations of their mines. Pedro Emilio Coll, the new Minister of Development, declared that as a result of this reform, and of the reorganization of the Servicio de Guardaminas,[12] there were no more 'obstacles preventing the acquisition and exploitation of concessions by either foreigners or natives'.[13]

In spite of the incongruities of some offending clauses in the mining laws, there were tangible results to Gómez's mining policies. Soon after it was announced in 1909 that the law was going to be reformed Dr Clodomiro Contreras, the owner of several gold-mining titles, left for London to negotiate his titles with British capitalists.[14] On 26 February 1910 the Venezuelan Syndicate Ltd was formed to take out an option to purchase Contreras' titles. In June 1910 Charles Freeman obtained from Pedro María Mata a 25-year contract to exploit and export the magnesium deposits on Margarita Island. A further success was achieved when agreement was reached in August 1910 between the government and the Pan American Iron Corp. for the latter to exploit the extensive iron-ore deposits at Imataca, Bolívar State. However, since the company was unable to utilize the concession, on 14 August 1911, after secret negotiations with Gómez, it was taken up by the Canadian Venezuelan Ore Co. Ltd, a subsidiary of the Dominion Steel Co. of Halifax (Canada). There was also renewed interest in the asphalt deposits of the country. On 4 December 1909 E. Stanley Simons took over the asphalt mines at Inciarte in Zulia and all the equipment belonging to the United States and Venezuela Co., which had previously worked the mine. The contract lapsed, and on 11 February 1911 General Juan María García took it over and later transferred it to Frank H. Phippen, who formed the Zulia Asphalt Co. to develop the property.[15] But it was the interest in the oil resources of the country that had the greatest effect on the government and on the economy as a whole in the ensuing years.

INTEREST IN OIL

The General Asphalt Co., which through its subsidiary the New York and Bermúdez Co. exploited the Guanoco asphalt deposits in eastern Venezuela, was interested in diversifying into oil production. As a result in May 1910, Lewis J. Proctor, Manager of the Pitch Lake Co. of Trinidad (another subsidiary of the General Asphalt Co.), arrived in Caracas to negotiate an oil concession.[16] Through Rafael Max. Valladares, a junior partner in the Bance law firm, a concession in eastern Venezuela was secured on 12 July 1910, and four days later it was transferred to the Bermúdez Co., yet another subsidiary of the General Asphalt Co.

On 16 November 1909 John Allen Tregelles had purchased for £500 a report concerning the oil resources of the country from General César Vicentini.[17] Through the 'active influence and support'[18] of Vicentini, who was a Deputy for Bolívar State in Congress, and Dr Nicomedes Zuloaga, Tregelles, together with N.G. Burgh, obtained on 10 December 1909 a concession covering over 3,000 square miles and extending over 12 of the 20 states that formed the Republic. Four days later Tregelles transferred his share of the concession to the Venezuelan Development Co., a British company, which in March 1911 conveyed it to the Venezuelan Oilfields Exploration Co. Ltd (VOC) for £48,334.[19]

The government was delighted with these mining activities. Aquiles Iturbe, the Minister of Development, in his Annual Report for 1911, drew attention to the future potential of Venezuela's mineral resources when he spoke of the promising mining activity which had started 'in some of our important mining regions due to the iron and oil exploitation contracts signed recently'.[20] The British company, however, ran into trouble when it applied for an extension of the exploration time allowed by the law. The government refused to grant this on the grounds that the company lacked the financial resources to develop its vast concession properly. The real reason for this refusal was that the General Asphalt Co. had taken steps to 'prevent its extension for the additional year'[21] so that it might itself acquire a similar concession covering the same area when the Tregelles concession lapsed in 1912. Rafael Max. Valladares, 'who is reported to be quite popular in certain circles in Caracas and to be capable of getting favours from the Government',[22] was to act once again as the intermediary between the government and the General Asphalt Co. According to the American Chargé d'Affaires negotiations were kept secret from the Cabinet because the country distrusted business dealings with American financiers. However, the 'knowledge that the firm is both wealthy and serious'[23] convinced Gómez to support the scheme. Proof of this was the transfer fee of Bs.400,000[24] which the company allegedly paid Valladares. This was, however, a ploy devised by the company to dupe the government into believing it indeed had sufficient resources to develop the concession. By his own admission Valladares did not receive any additional payment over and above the ordinary legal fees which he charged as the company's lawyer 'complying with instructions'[25] received from Proctor.[26] It is reasonable to assume that this is correct as the Bance law firm had always maintained a close relationship with the General Asphalt Co. through its defence of the subsidiary New York and Bermúdez Co. during its long legal battle with the Castro Administration. At first both the Minister of Development and the Committee of the Council of State appointed to review the contract opposed the scheme until Gómez 'put pressure on them to force them to consent'.[27] Thus on 2 January 1912 the Council of State approved the Valladares contract. The concession granted to him was identical to that previously held by Tregelles, and was to pay Bs.1 per hectare per annum surface tax, and Bs.2 per ton of oil extracted. Two days later, on 4 January, a government decree transferred the concession to the Caribbean Petroleum Co. Ltd (CPC), a further subsidiary of the General Asphalt Co. The General Asphalt Co. had also acquired from Andrés J. Vigas in June 1913 a concession covering the Colon District of Zulia State and had formed the Colon Development Co. (CDC) to exploit it.

The General Asphalt Co., however, ran into financial difficulties and, unable to dispose of its two concessions in the U.S., sent James Clark Curtin to Britain to negotiate with the Royal Dutch-Shell Group (Shell). Henri Deterding, the head of Shell, was extremely impressed by the geological report prepared by Ralph Arnold and decided to acquire both the Valladares and the Vigas concessions. He later described the transaction to acquire CPC as 'our most colossal deal'.[28] The entrance of Shell meant that CPC was provided with 'ample finances, techniques and experience so badly needed in a project of this size'.[29]

The Venezuelan government remained unaware that Shell had acquired these two concessions, but it could not help but notice that the country was beginning to attract more attention, especially among British capitalists. At this time also a friend of Gómez had been successful in interesting a group of British financiers in acquiring his concession. On 28 February 1907 General Antonio Aranguren, with the help of General Francisco Antonio Colmenares Pacheco (Gómez's brother-in-law), whom Aranguren had helped with business deals in Trinidad, obtained an extremely valuable concession to exploit the oil and asphalt deposits of the Maracaibo and Bolívar Districts of Zulia State for 50 years. Aranguren, unable to find anybody interested in his concession, travelled to London in 1910 in the hope of getting British capitalists to develop it. Acting through his solicitor, Edward O. Goss, he was able to get Duncan Elliott Alves to take out an option to purchase the concession.[30] As a result, on 8 April 1913 the Bolivar Concessions Ltd, with an authorized capital of £30,000, was registered in London to acquire Alves' option on the Aranguren concession. On 22 May 1913 that part of the Aranguren concession relating to oil was sold for £35,000 in cash and £70,000 in fully paid £1 shares to the Venezuelan Oil Concessions Ltd, a new company in which Alves and Aranguren were shareholders, the Bolivar Concessions Ltd retaining the asphalt–bitumen interest in the concession.

There was great hope for the Company's success in Venezuela. The President of Zulia stated that he would declare a public holiday on the day oil was discovered, and Pedro Emilio Coll, the Minister of Development, declared in his Annual Report for 1913 that the country's oil resources 'had ceased to be treasure hidden in the entrails of Venezuelan soil, to be revealed on the surface . . . [and] that the exploitation of our oilfields will surely be a further attraction exercised by the country in the field of commercial enterprise'.[31] Rapid progress was not smooth, however. The generally unhealthy working conditions, faulty equipment, and legal problems about the company's concessions served to diminish the rate of work. The company was also running out of finance and was eventually taken over by Shell.

The early history of the Venezuelan oil industry shows the difficulty encountered by the oil concessionaires in attracting foreign capital. Only after considerable time and effort were they able to transfer their oil concessions to foreign companies. But the problem did not end there, because the high cost of development, due to an inadequate infrastructure, meant that the companies had to seek outside technical and financial assistance. The entry, therefore, of Shell with its ample resources was not only welcomed by the oil concessionaires but by the Gómez Administration too, which had tried since the December 1908 coup to encourage the development of the mining and oil industries. The entry of Shell into Venezuela assured the development of the country's oil resources, and at the same time redounded to its own advantage. Although Deterding's decision to develop the country's oil resources was a bold step at the time, the advantages associated with being first (such as securing the best oil-bearing lands and favourable taxation) gave Shell a considerable edge over its rivals. By the early 1920s Shell, with its three operating companies, was poised for a massive increase in oil production, which would push the country into the forefront of the major oil producers of the world.

THE GOVERNMENT BECOMES MORE INTERESTED IN OIL

The interest in oil which occurred during 1913 increased the government's awareness of the value of this potentially large source of revenue. In consequence the contracts negotiated in early 1914 were more onerous for the oil companies than had previously been the case. For instance, the Val de Travers Asphalt Paving Co. Ltd, which in 1904 had acquired seven asphalt mines of 300 hectares each in Monagas State[32] from the Compagnie des Asphaltes de France, wanted to adapt these titles to oil exploitation. The committee appointed also felt that the tax of Bs.2 per ton of oil produced 'did not hold any relation with the price and importance attained by the oil'.[33] The proposed contract was rejected, and the committee concluded that it 'would be prudent to postpone all leases of oil mines until the country enacted a law which corresponded to these circumstances'.[34] Further proof of the government's harder line was the rejection of the proposed contract by the Venezuelan Petroleum Exploration Co. Ltd, a Shell subsidiary which had been formed and registered in London on 13 November 1913. A.J. Van Oosteveen was sent out to negotiate some oil properties with the government. After consulting Dr José Gil Fortoul, the Provisional President, and Pedro Emilio Coll, the Minister of Development, in December 1913, he presented a contract to the government[35] which it later rejected because it found the proposed initial exploration period (four years) and size of the production blocks (500 hectares) unacceptable. A further sign of this greater awareness of the oil potential of the country was a government decree of 19 September 1914 which declared inalienable all petroleum, asphalt, and any other similar substances for which concessions had until then not been awarded. The government would take over the direct administration of such properties. As a result of this resolution, on 24 September 1914 the government took over the administration of the coal mines at Naricual, Capirical, and Tocoropo, and of the port of Guanta, which served the coal mines of Anzoátegui. Similarly, when the contract belonging to General León Jurado, the President of Falcón State, to develop the coal deposits there expired, the government took over their management, and also that of the La Vela Coro Railway.[36] The management of these coal mines was the first attempt by the government to administer and develop an important natural resource.

With the enactment of the new 1914 Constitution and the decree of 19 September 1914, it became necessary to promulgate a new mining law to bring the law into line with these developments. The 1914 Constitution increased the political power of the Executive, and also abolished the Council of State, thus giving the Executive a freer hand in all its negotiations for mining contracts. To counterbalance this, article 58, which stated that Congress would now have to approve all mining concessions and other contracts made by the government, was inserted in the new Constitution. The new Mining Bill presented by Santiago Fontiveros, the then Minister of Development, to Congress on 11 June 1915 incorporated the inalienability of oil and asphalt mines, and the government's right to administer them, 'with the right to increase taxes'.[37] All this was incorporated into the new law which was passed on 26 June, with the one exception that concessionaires would be exempted from any future taxes levied. The law marked the beginning of the separation between the mining and oil industries which was achieved in 1920 when the first Oil Law was enacted. The new law also allowed the Executive the freedom to negotiate contracts, but within the limits laid down by the law.

EFFORTS TO INCREASE PRODUCTION

The Venezuelan government was aware of the increase in oil activity which was taking place, and the seriousness with which some of the companies were taking their work, but revenue from this source was low. For example, in 1914 CPC only had five wells in production out of the 1,028 blocks selected. Consequently, in the government's view, the production of crude oil was not in proportion to the great extent of the concession held, and this was having a detrimental effect on government revenue. Santiago Fontiveros, the Minister of Development, therefore refused to accept the tax payments submitted by the company 'as long as the annual production of 500 tons of oil, equivalent to the minimum of one thousand bolívars (Bs. 1,000) the company had to pay, had not been verified'.[38] The reason for the government stipulating a minimum production rate per block held was, as Fontiveros pointed out, to 'increase employment in the mining sector'.[39] A solution was reached, following the government's threat to rescind CPC's concessionary blocks, when the company agreed to pay a minimum tax of Bs. 1,000 per mine and to 'employ at least 20 workers (half having to be Venezuelans) for each well drilled every working day of the week'.[40] Nevertheless, within the Mining Department of the Development Ministry, there was a growing resentment of the company's 'idleness and dirty tricks'.[41] Horacio Castro, the Director of the Mining Department, felt that Dr Grisanti, one of the Ministry's legal advisers, was 'notoriously biased'[42] in favour of the company, and that he had helped it to obtain 'from the Development Committee of Congress a file on common land which it did not want approved, and which later *appeared* in *this* Ministry without the survey plan and pages detailing the Caribbean's opposition'.[43] Furthermore, CPC *'has caused to vanish* from this Department notes'[44] (belonging to Gómez) 'which would prove embarrassing if its adversaries exhibited them at a trial'.[45] Castro therefore felt that it was time to 'free Venezuela's treasures from the greedy claims of the foreigner',[46] and, when the company applied in 1916 for an extension to its three-year exploration period (because, it alleged, its operations had been paralyzed due to the political unrest of the country), Castro advised the Cabinet not to approve the extension as 'millions are being gambled in this bet'.[47] His advice did not go unheeded: the Cabinet refused to allow CPC an extension, arguing that the company had had sufficient time in which to explore. A compromise was reached whereby the company would renew its titles by paying a fine of Bs.1,000 per title renewed. By paying a fine of Bs.409,000 (which together with stamp duties brought the total amount raised to Bs.500,000) the company retained 409 blocks out of the original 1,028. This was a large amount for any company to pay, and it showed the seriousness with which the company took its concession. But in 1917 the company faced the threat of the annulment of its contract by the government when on 16 February E.W. Hodge, a black Trinidadian, entered a complaint at the Ministries of Finance and Development, claiming that the Valladares concession was illegal and unconstitutional.[48] Under the Código de Hacienda of the time all property which was subsequently found to have been granted under illegal terms reverted back to the state, and the claimant acquired two-fifths of it. An American company interested in oil developments, the Paria Transport Corporation, acquired Hodge's claim for Bs.10,000 on 23 February 1917.[49] Hodge's claim, as Alejandro Urbaneja, the Attorney-General, confirmed, was a very good one,[50] but as he also pointed out, there were other considerations at stake since the company could claim compensation from the government for having approved an uncertain contract. There was also the threat of intervention by foreign governments on the company's behalf. It was thus advisable to follow a discretionary policy which would minimize the chance of this occurring in the future.[51] This view prevailed at the Cabinet meeting on 17 March 1917 and, after Gómez had been consulted,[52] on 22 March 1917 the Hodge claim was considered 'inadmissible'.[53]

EARLY LINKS BETWEEN GOMEZ AND THE OIL INDUSTRY

This greater awareness of the country's oil potential had the pernicious effect of increasing the corruption and intrigue amongst Gómez's family and entourage, the consequences of which would be felt up to 1935. At the forefront of this were José Vicente Gómez Bello, Gómez's son; Juan Crisóstomo Gómez (Juancho), Gómez's brother; and Julio F. Méndez and Carlos Delfino, Gómez's sons-in-law (married to his daughters, Graciela and Josefa respectively).

In 1914 Colonel José Vicente Gómez was appointed a member of the Comisión Permanente de Fomento of the Chamber of Deputies in Congress,[54] and the following year Carlos Delfino also joined the committee.[55] This was the most important congressional committee for the mining industry, because it vetted all the contracts which were entered into by the government and private individuals, and then recommended to Congress which contracts should be approved.[56] José Vicente and Carlos Delfino were therefore in a privileged position to gain valuable information about the country's mining development. Although this information was used for personal gain it did have one advantage for the country as a whole, for now the development and monitoring of the country's oil and mining industries would be linked directly to the personal gain of Gómez's family, thus ensuring that the head of the country was intimately informed on the progress and problems of the industry. Other government officials were similarly involved with the oil industry, for although they were strictly forbidden by law[57] to hold oil concessions themselves, they were nevertheless able to obtain these concessions by means of nominees.

The close involvement of the Gómez family with the oil

industry can be seen from the following examples. In 1916 Felipe Alvárez Cienfuegos, who carried out a geological survey and exploration of Zulia between 1914 and 1916, informed Gómez 'of the land with oil-bearing potential which could still be acquired'.[58] Gómez suggested that Alvárez carry out a more detailed survey of the area between the rivers Santa Ana and Lora in Zulia 'to determine fully the extent of the oil-bearing rock and its exploration potential'.[59] On his return to Maracay in 1917, having accomplished his task, Alvárez was unable to see Gómez. Nevertheless, a few days later he received a visit from Rafael Requena and Addison H. McKay (who had arrived in Venezuela to acquire the copper mines owned by the South American Copper Syndicate in Yaracuy State, an affair being handled by Requena), who informed him of Juancho Gómez's request that he 'hand over all the facts relating to the study, while at the same time requesting my help for a bit longer, all this in exchange for a document in which I was given 25 per cent of the profits'.[60] This information was later used by Julio F. Méndez and McKay to acquire some very lucrative contracts in 1919. Gómez also commissioned Pedro Guzmán, who had represented Alvárez in 1916, to make a survey of the oil deposits of Lake Maracaibo and its shore. Alvárez carried out this survey as well, and handed over the results, for which he was paid Bs.20,000, to two friends of Gómez.[61] In addition, Juancho Gómez was closely associated with two legal cases brought against VOC and CPC. In the latter case Messrs Valbuena, Espina and Bohórquez claimed damages from the company because part of the Valladares concession covered three asphalt mines awarded to them in 1904. Juancho was interested in obtaining a favourable settlement for Valbuena, Espina and Bohórquez because he had acquired 25 per cent of their property. In spite of Juancho's influence on 15 April 1916 the court decided in favour of the company.[62] In the case of VOC, Lorenzo Mercado sought to establish his claim to a quarter of Aranguren's concession.[63] Mercado tried to use his influence with Juancho Gómez to gain a favourable sentence from the Superior Court. During the trial Dr Delgado Briceño, the Secretary-General of the Federal District Government (which was headed by Juancho Gómez) and a man 'unscrupulous and brutal in the extreme',[64] saw Drs Juan Pablo Colmenares,

Juvenal Anzola and C.J. Rojas, who were judging the case, and told them that he had received direct orders from Juancho Gómez that 'the decision in the aforementioned trial should be favourable to Mercado'.[65] A few days later Rojas, in his capacity as Chancellor to the Superior Court, was called to the Court-Secretary's chambers where Briceño informed him that the court must decide in favour of Mercado.[66] The court, however, did not bend to this pressure, and decided on 28 June 1917 in favour of Aranguren and the company.[67]

José Vicente Gómez was also involved at this early stage with oil concessions. In 1916 he secured from Santiago Fontiveros, at that time the Minister of Development, an oil and coal concession in Zulia for Arístides Soto Bracho of Maracaibo.[68] In late 1917 he sent General Pablo Romero Durán to see Gumersindo Torres, the then Minister of Development, in order for him to 'study several files, because I am interested to know how far several mining applications have progressed'.[69] José Vicente himself owned two iron mines (Venezuela 1 and 2) in the Municipality Antonio Díaz, Delta Amacuro Territory.[70]

Interest in the oil industry within Gómez's inner circle increased further after the industry had been clearly established on a permanent basis in the 1920s. This awareness, which was motivated primarily for personal gain, had the effect of linking Gómez's interests in developing the country's natural resources even closer with his own personal interests. He was therefore acutely aware of the developments occurring in the oil industry, and had strong personal and political reasons for welcoming the growth and prosperity of the industry.

INTERNATIONAL INTEREST IN VENEZUELAN OIL GROWS

During these years foreign interest in the country's oil resources increased further. In November 1916 Bulners Tavares, representing V.V. Lebedjeff Engineering and Supply Co., proposed to Gómez that the government together with the company should exploit the country's oil resources, something which the government rejected.[71] Alves, who was closely connected with VOC, formed a new company in London on 29 November 1917 called Bolivar Concessions (1917) Ltd to acquire from Bolivar Concessions Ltd the Planas concession for £90,000 in fully paid

£1 shares. A conditional agreement was entered into for the sale of part of the assets to Messrs. Sperling and Co., and on 15 January 1918 the British Controlled Oilfields Ltd was organized in Montreal (Canada) with a capital of $40 million. On 15 September 1917 CPC presented Gómez with the first drum of petrol produced by the San Lorenzo refinery. The *Nuevo Diario* newspaper reflected the mood of anticipation prevalent in the country over the future of the oil industry when it stated that 'it is today's hint of an even richer future in store'.[72]

The increased activity in the oil industry resulted in the government seeking greater control and supervision, as well as modifying the law to allow concessions to be granted more swiftly. There was a need to organize a more efficient way of dealing with the applications for mining contracts in the Ministry of Development. This would eventually lead to the enactment of legislation dealing exclusively with the oil industry.

2
The legal framework

As early as 1915 the government was already considering how to improve its control over an oil industry which was still very small. In his annual report for 1916 Manuel Díaz Rodríguez, then the Minister of Development, announced that the first task towards achieving this end was to set up a proper filing system from which the mining titles and all related information could be easily retrieved.[1] He also saw the need, already suggested by the Council of State in 1914, to separate the oil industry and the coal mines from the mining industry as a whole. The present mining legislation had been drawn up as the best way to manage gold and copper mines, but the law did not adequately cover the extraction of oil. In drawing up new legislation Díaz Rodríguez stressed the need to specify the characteristics peculiar to each mining group. Moreover, the Minister alluded to the tortuous and disorganized manner in which mining titles were obtained, and the great delay caused by Congress having to approve them. It was necessary to streamline the whole procedure by allowing the Executive to award concessions without congressional approval. This did not mean that the Executive would be given a free hand because, Díaz Rodríguez argued, the award of an oil concession was merely an administrative measure since the terms of the contract were given in the law itself. As no negotiation of the terms took place, congressional approval was unnecessary because it did not increase any safeguards.[2]

A change in the rigid taxation system was also envisaged, especially as it did not take into account the international price fluctuation of the minerals in question. Because of this, the Minister wanted to introduce an *ad valorem* tax on mineral production.[3] He also felt that surface tax should be charged

21

from the time the mining title was granted and not from the date of production. The reason was that only one per cent of all the mining titles awarded went into regular commercial production, and this naturally had a detrimental effect on government revenues; moreover, it would also stimulate the mining industry from the lethargy into which it had slipped.

There was also an urgent need for more rigorous supervision of the industry. Many mine owners obtained their mining titles not to develop them but to enter into speculative deals with foreign capitalists, 'relegating to fortuitous circumstances what constitutes the real object of the law: effective exploitation'.[4] A further example of this was that the mine owner who did not operate the mine could obtain a three-year extension to his title by feigning production when the mine inspector came to check the operation of the mine. Once this was obtained, work on the mine ceased because the owner was only interested in retaining his mining rights in order to transfer them to a foreign enterprise. The Ministry's inspection service was also deficient, and there was an urgent need to train more efficient *guardaminas*. Díaz Rodríguez, however, did not see his efforts to fruition, for international events intervened. Ever since 1917, when Gómez had refused to follow the U.S. into the First World War, relations between the U.S. and Venezuela had left much to be desired. From the outset of the conflict Gómez had declared a strict policy of neutrality, but in order to appease the Allies, who had labelled him pro-German, he decided to dismiss the most pro-German of his Ministers, Pedro Manuel Arcaya, the Minister of the Interior, Manuel Díaz Rodríguez, the Minister of Development and Carlos Aristimuño Coll, the Minister of Education, replacing them with Ignacio Andrade, Gumersindo Torres, and R. González Rincones respectively.[5]

THE 1918 OIL REGULATIONS

As we have seen, the government was moving towards a stricter control of the mining industry, and in particular of the oil industry. For this reason, when Torres on 12 November 1917 presented 30 contracts for the Cabinet's approval, they were all rejected because 'the benefits and advantages offered were inferior to the Caribbean contract, which in turn was considered very onerous to the country'.[6] The Cabinet decided to postpone the approval of all oil contracts until it issued a set of regulations which 'would serve as the basis for Congress in its next session to enact a law covering the industry which constitutes one of the most important and lucrative sources of revenue and prosperity for the Nation'.[7] Torres, who continued his predecessor's reorganization of the Ministry of Development immediately set about the task of drawing up a new *Reglamento* for the oil industry. He was determined to produce a set of regulations that would be beneficial to both parties. He obtained as much information as possible from other oil-producing countries so that the *Reglamento* enacted would be the result of 'having complete knowledge of the facts needed to act sensibly, thus preventing any criticism by future generations that we did not know how to safeguard our natural resources'.[8] Like his predecessor, Torres was very much aware of the country's oil potential, and wanted to amend the mining law to take into account the special needs of the oil industry. However, he also stressed the need for care and caution when amending the law. He drew attention to the fact that 'until recently, oil exploration and exploitation contracts were awarded thoughtlessly, consequently the country had obtained little or no benefits at all from them'.[9] He rejected the idea that a full oil law should be enacted, but recommended that 'instead of jeopardizing the future by subjecting the industry to the rigid norms of an imperfect law, that regulations be issued cautiously, amending them whenever the need arises'.[10] Thus the Executive should have 'ample regulatory powers in oil matters'.[11] Like Díaz Rodríguez, Torres also suggested that oil and other minerals should be taxed on an *ad valorem* basis on the price which the mineral fetched in the open market. This would have the advantage of being both simple to collect and fair, for the government's revenue would be a fixed percentage of the mineral's price in world markets. Torres also saw the need for establishing a minimum annual export tax, 'even in cases where no minerals had been extracted after the end of the period allowed to start exploiting the mine',[12] in order to stimulate production. A further stimulus would be provided by permitting the reacquisition of a mining title which had lapsed. Up to then, once a mining title had been rescinded its reacquisition was a lengthy procedure. By payment of a renovation fee, the matter would be simplified. Finally, Torres recommended that Congress grant his Ministry an increase in its annual budget in order to create a new Dirección de Minas within the Ministry.[13]

These points were discussed on 22 March 1918 at the Cabinet meeting to consider 'the Bills to be presented to Congress following the instructions'[14] received from Gómez. It was agreed that a new mining law would be enacted with two separate *Reglamentos* to cover oil and mines. Torres' bill included a very important and new provision which had been suggested by Gómez himself, which was that National Reserves, comprising half a company's exploration area, should be created.[15] Torres considered the National Reserves to be a 'guarantee for the future'[16] because the government would acquire lands which would increase in value once oil had been found in the adjoining blocks. These could then be auctioned to the highest bidder, and the government would obtain greater benefits without the burden of conducting an expensive geological survey to determine oil-bearing lands.[17]

These ideas were all incorporated in the new mining law, enacted on 27 June 1918, which for the first time separated oil exploitation from the rest of mining. The enactment on 9 October 1918 of a separate *Reglamento* for oil and similar substances was official recognition of the importance that oil and the expected revenue from its exploitation were assuming in the country. The new law established much more onerous production terms for the companies; for example, concessions would run for 30 years, and the maximum exploration concession would be only 15,000 hectares, with exploitation blocks reduced to 20 hectares. Surface tax would be an initial Bs.0.05–0.10 per hectare per annum, and then Bs.2–5 per hectare per annum, 'according to the richness of the zone and its situation',[18] and exploitation tax would be 8–15 per cent of the 'commercial value of the mineral'.[19]

Soon after this, on 1 July 1918, the Dirección de Minas was created within the Ministry[20] to deal with the new era of oil exploitation which Torres was convinced was about to begin, especially since enormous quantities of oil would be needed for reconstruction after the end of the European conflict. However, contrary to these expectations, oil production did not take off that year, and on 6 March the government was forced to re-establish the tax on locally refined oil which had been abolished the previous year.[21]

STIMULATION OF OIL PRODUCTION

The Shell group of companies, which were the only ones in operation, were seen to be lagging behind in production. There was an urgent need to stimulate production. On 7 February 1918 Torres informed CPC that he did not consider the company's concession to be working at full capacity.[22] The company disagreed with this assessment, and felt that it was being harassed. In a long memorandum sent to Torres on 12 April Proctor, CPC's manager, set out the argument to substantiate this suspicion. He believed, first of all, that there had been a misunderstanding at some point because the government's policy had always been to encourage and help the company, and therefore 'the opposition the company experienced could only be the result of a misunderstanding or erroneous information'.[23] Proctor declared that his company had developed the country's oil industry, and had brought in machinery valued at Bs.4 million.[24] The company's total investments, amounting to Bs.20,782,842,[25] were the largest in the mining industry; for example, total investment for the Bermúdez Co. amounted to Bs.4,319,820;[26] for VOC, BS.2,316,996;[27] and for CDC, Bs.3,750,000.[28] Only the El Callao Gold Mining Co, with a total investment of Bs.20 million, compared favourably with CPC.[29] He further alluded to the fact that his company's concession was smaller than the four large concessions granted in 1907,[30] and that the company had developed its concession more quickly than had the other companies operating in Venezuela. The company's contract was stricter than those awarded to others; for example, the company had to drill on each of its selected blocks, whereas in other concessions, which in addition granted the concessionaire the exclusive right to all the oil produced, it was necessary to start the exploitation of only one site of the contract within the specified time set for exploration.[31] Proctor also asserted that the company paid more taxes than the other companies,[32] and answered the criticism that his company was not working its concession to the full by claiming that the first wells drilled were the ones which were now producing oil. (These were drilled by a lightweight rig which the *guardaminas* did not consider as proof of effective production.)[33] The company further argued that, under article 45 of the 1910 Mining Law (under which the contract had been signed), for a mine to be considered active there had to be at least five workers and 'some machine in activity',[34] but that under article 45 of the 1918 Mining Law the minimum requirement had been changed to ten workers and 'the necessary machinery'.[35] Proctor therefore pointed out that the law did not 'specify the type or size of machinery, depth to which the mines must be worked, or the time over which work must take place at the mine'.[36] Furthermore, according to article 44 of both mining laws, the concessionaire was allowed to suspend work on his mines for a maximum of three years. This right the company had exercised on certain wells. The *guardaminas* had been misled when they arrived at a rig and could not detect any activity; furthermore, the company had informed the Ministry whenever it had started operations. It was thus up to the Ministry to advise the *guardaminas* when to check the company's work 'and not, as was the case, months later, when for various reasons, and making use of the right already alluded to, the company suspended work on some of the blocks, with the consequent effect that the *guardaminas* found them at a standstill'.[37] Torres sent the whole question for discussion at Cabinet-level, and a decision was reached whereby the company paid the minimum tax of Bs.1,000 for each of its 185 blocks which were now declared to be in exploitation. It was also decided that the company would place its other blocks (235) under exploitation within the next three years, and that the company would pay Bs.1,000 exploitation tax on each of these blocks.

A more serious clash developed between CDC and the government, for this company had only selected three blocks for exploitation amounting to a mere 800 hectares, out of its colossal concession of 1.5 million hectares. The government observed that the company was exploring its concession on much too small a scale so that 'they could not claim exclusive rights to all petroleum etc. deposits in the district of Colon – and the Minister would declare the contract as having lapsed'.[38] The company, on the other hand, argued that its concession was just one huge mine, and having established this, it was only willing to pay taxes, amounting to Bs.1,200, on its three selected blocks, instead of the Bs.3,800,000 demanded by the government. In early 1918 Torres warned Proctor, the company's manager, of the need to 'put right the irregular, not to say illegal, situation of the Colon claiming possession of the District when it has not fulfilled the clause in its contract which calls for it to demarcate deposits and exploit them'.[39] Torres wanted to see the nation gain 'the benefits which the fulfilment of that clause implies'.[40] Shell, however, was unwilling to develop the property until it had solved the problem of Vigas' vendor's rights,[41] which the company deemed to be disadvantageous to it. This led to a very delicate situation later on when the American oil companies started to look at Venezuela as a profitable investment proposition.

Far from driving the oil companies away, the government's aim was to get them to produce more oil, and thus to develop their extensive concessions. It was realized that without foreign companies the oil resources of the country would not be developed. But foreign capital needed to have confidence in the country's rulers in order to risk investment. British capitalists, who up to then had financed the development of Venezuela's oil resources, were seen as an important source of foreign capital. For this reason, when Alves and Pedro César Domínici informed Gómez that Mercado's suit against Aranguren and VOC was causing 'great dissatisfaction'[42] among the shareholders, to the detriment of Venezuela's credit-worthiness in the City of London, Gómez immediately replied assuring Alves and his shareholders that a quick solution to the dispute would be reached. Gómez managed to bring together both parties in the affair, and a private settlement outside the courts was effected. Moreover, Dr José Santiago Fontiveros, who had been the company's attorney in the Mercado case, was appointed Commercial Attaché at the London Venezuelan Legation in August 1919.

THE 1920 OIL LAW

It became apparent that stable legal conditions under which foreign oil companies could operate needed to be established.

Torres' commendable idea of adapting the existing regulations to the changing conditions of the oil industry had the drawback of fostering the belief among oil operators that their operating conditions were unstable. For this reason it was necessary to enact an oil law to cover oil development, and to give the operating companies the security they wanted. In drafting the new law Torres was most interested in comparing the relevant legislation of other countries in order to provide an oil law that would be both acceptable to the oil companies and competitive with the other oil-producing countries. He also checked the real commercial value at Venezuelan ports of the Bs.2 per ton exploitation tax on oil which the Shell group of companies paid. The company which Torres consulted reported that the value of one ton of oil at their well-head was Bs.24.55, which meant that the Bs.2 per ton tax was equivalent to 8.2 per cent of the oil's value.[43] He was also considering the formation of a new Dirección General de Administración 'to centralize and control all sources of revenue'.[44] In December he requested Rafael Hidalgo Hernández, a leading Venezuelan lawyer in Paris (who was later involved in drafting the first three oil laws), to make a comparative study of the Venezuelan dispositions concerning oil and those of the oil-producing countries to strengthen 'the notion which I suspect is correct that our measures secure the oil future of Venezuela, and that our oil will not be exploited as, in the past, without benefitting the country'.[45] He also requested Vicente Lecuna, just before his trip to Washington to attend the Second Pan American Financial Conference, to make a comparative study of 'the country's oil and coal mining dispositions with those of the U.S. and Britain, with special reference to taxation'.[46] Torres also commissioned his close friend, Pedro Manual Arcaya, to recommend the most efficient way of stimulating production and increasing government revenues. Arcaya felt that there were two roads open to the government: it could either exploit the oil deposits directly, or grant concessions to a large number of people. Arcaya dismissed the first solution as being too expensive, with the added burden of Venezuela having to market its oil through the large oil companies.[47] The only real solution Arcaya advised would be for the government to act as an overseer, handing out concessions to companies which would pay taxes for the privilege of exploiting the oil deposits. The easiest and most efficient way of doing this, if the country had a detailed geological survey, would have been to award oil exploitation blocks to the highest bidders. However, the drawback here, as Arcaya pointed out, was that the large oil companies would not be interested in such a system, with the result that only small speculative companies would take up the blocks, and 'thus tenders which bettered the Valladares contract would not be submitted'.[48] The system up to 1918 had been for concessionaires to acquire large tracts of land at a relatively small cost, and then transfer them to companies at absurdly low prices.[49] The disadvantage for both the government and the nation as a whole was that the concessionaire was not the owner of the land which his concession covered, allowing him to sell the concession for 'sums which are large in relation to the amount invested by him, but which are ridiculously meagre in comparison to what the acquiring company would have to spend in any other country to obtain the right to exploit such a vast extension of land'.[50] However, when the company had to negotiate with individual landowners or buy up the land as in the U.S., then the acquisition of such vast concessions became too expensive. So Arcaya argued that it would be beneficial for all concerned if property owners were given the temporary right to develop the oil resources on their own land. The benefits of this system were first, that it would be more equitable to the property owners if the country turned out to be a large oil producer. Second, that property owners would not be deprived of their constitutional right to drill artesian wells. This had occurred in the Miranda District of Falcón State when property owners had tried to drill artesian wells, and CPC had opposed it until the property owners had signed a document 'which protected its [CPC's] right to explore the subsoil, and which it considered had preference over the rights of the landowners to drill wells'.[51] Third, that property values would increase because the oil companies would have to deal with property owners individually, and not with a single concessionaire. This would have a beneficial effect generally because the 'increase in value of the land would naturally increase the general prosperity of the region and government revenues'.[52] It was felt that by granting property owners the right to acquire the subsoil oil rights to their property, interest in the country's oil resources would be stimulated. It was acknowledged that the property owners could not possibly develop the holdings themselves, and so they would have to seek foreign capital, flooding New York and London with Venezuelan oil concessions. The oil companies would have to take an interest in Venezuela because they would be unable to determine whether their competitors had acquired valuable concessions in the country.[53] Thus, if only a small number of concessions were transferred, the government would have achieved its objective of attracting foreign capital to develop the country's oil resources. Consequently, U.S. oil companies received considerable attention from the various Venezuelan consulates, which mailed numerous pamphlets and held confidential discussions with the companies, and A.J. Woodward, a well-known engineer, was invited in 1919 by the Venezuelan government to determine the country's oil possibilities.[54]

A further advantage to this system, which Arcaya omitted to state, was that it would give the landowners, an important political element in the country, the opportunity of sharing in the potential oil wealth. Up to then this had not been possible because the granting of concessions was in the hands of the government, thus permitting those closely connected with the government to receive the most valuable concessions. As we have already seen, Juancho Gómez, Julio F. Méndez, and Addison H. McKay had cheated Alvárez Cienfuegos of his geological survey of parts of Zulia. Armed with this information Méndez was able to obtain in 1919 seven oil concessions in Zulia, and McKay, in the same year, secured six oil concessions in Falcón, and nine coal concessions in Zulia. Other people who were closely associated with Méndez, and who were probably front-men for Juancho Gómez, were Domingo M. Navarro, who obtained four concessions, Eduardo Navarro (five concessions), Rafael Font Carrera (six concessions), and M.A. Alvárez López Méndez (five concessions).[55] In July 1919 Méndez, who had travelled to New York together with McKay to negotiate the oil concessions, informed Gómez that he had a group interested in the 'oil deposits, and it is possible that a good contract will be signed'.[56] On 10 November 1919 McKay transferred six of his concessions to the Venezuelan Oilfields Co. Ltd,[57] a subsidiary of the Sun Oil Co. of the U.S. In the same year he negotiated twenty oil concessions, amounting to 300,000 hectares, with James S. Wroth, representative of the Maracaibo Oil Exploration Co., a company incorporated on 9 November 1919 under the laws of Delaware State with a capital of 250,000 shares of no par value, of which 150,000 shares were subscribed at $26 within a month.[58] In 1920 McKay further negotiated with the Sun Oil Co. five concessions located in Trujillo State which belonged to Juancho Gómez and Rafael Requena, but which for legal reasons were under the names of Félix R. Ambard and Héctor Finol.[59] Later in April 1920, McKay, representing both the Sun Oil Co. and the Standard Oil Company (New Jersey) (Exxon), submitted an offer in writing to the Ministry of Development of $1,350,000, and an equal sum to Gómez, for the privilege of exploring for one year the property belonging to CDC, VOC, and the British Controlled Oilfields Ltd, which were at that time threatened by the Venezuelan government with annulment of their respective concessions for breach of contract. Other members of Gómez's family were equally interested in dealing with oil concessions. On 17 March 1920 Francisco G. Travieso, who had been warmly recommended to Torres by José Vicente Gómez,[60] obtained an oil concession covering 71,000 hectares of the Colina District of Zulia State. There had been four other bidders for the concession, namely Rafael Cayama Martínez, Ramón Ayala, Ignacio Andrade, and Juan Augusto Perea, but 'Travieso's proposals over most of the District were the best, except in a small sector, Macornica, where Perea offered fifteen per cent instead of the ten per cent which Travieso proposed.'[61] Gómez agreed with this, and thus Perea obtained 4,000 hectares in addition to Travieso's 71,000 hectares. In the same year General José María García transferred four concessions he had acquired in Zulia and Trujillo States to the National Mining Corporation Ltd.[62] Santos Matute Gómez, President of Zulia at the time and Gómez's cousin, also requested Torres to grant him several concessions in Zulia for which he had sent the necessary papers to Juancho Gómez.[63] Juancho, too, recommended to Torres that Esteban Ramón Paris, of the Maracaibo trading house of Juan E. Paris, be granted several concessions and also permission to establish an oil refinery in the country.[64]

It would be wrong to assume that this abuse of power only

occurred among the Gómez family members who were in the government, for other functionaries sought to gain oil concessions with equal eagerness. On 10 September 1919 Manuel Antonio and Luis Sira presented their proposals for oil concessions in the Delta Amacuro Territory to the Ministry of Development. The Sira brothers were nominees for Victoriano Márquez Bustillos, the Provisional President of the country, as Elías Rodríguez, Secretary-General to Márquez Bustillos explained. He told Torres that the proposals 'belong to doctor Márquez Bustillos who would be grateful if doctor Torres would approve them in preference to any others as long as this is legal, and if there is a better proposal then doctor Márquez Bustillos is willing to better it'.[65] At the same time people outside the government sought to curry favour with Gómez, presenting him interesting business proposals. B.G. Maldonado, for example, requested Urdaneta Maya, Secretary-General to Gómez, to recommend to Gómez his proposal which involved an American company.[66]

Arcaya, therefore, recommended to Torres that exploration zones should be a maximum of 1,000 hectares, that oil should be taxed at a percentage of its commercial value, that excess profits over 12 per cent should be subject to a surtax of 5–20 per cent, and that property owners should be able to obtain within a specified period (possibly one year) the oil rights to their land.[67]

The international oil situation had changed radically after the First World War with both the American and British governments engaged in actively promoting their oil companies to seek new oil supplies. The logistic problems faced by Middle East producers, and the proximity of Venezuela's oilfields and their relative ease of access placed the country in a favourable position. In addition Colombia, which was viewed by the oil fraternity as a much more promising producer, found that the enactment of its 1919 Oil Law scared many prospective oil companies away.[68] This left Venezuela in a very good position, if it could offer favourable operating terms and the stability needed by the oil companies. Moreover, Shell's considerable presence in the country acted as an indirect guarantee that the oil resources of the country were large.

In Venezuela, too, there was an increased awareness in commercial circles of the value of oil and what this signified for the country's progress. In 1919 the tremendous share price increase on the New York Stock Exchange experienced by the General Asphalt Co. and the Carib Syndicate (companies associated with oil developments in Venezuela and Colombia) did not go unnoticed. The prestigious *Boletín de la Cámara de Comercio de Caracas (BCCC)* pointed out in an editorial in 1920 two reasons for this: one was the existence of large oil deposits in the country, and the other was the fear of a scarcity of oil.[69] Further confirmation of the increasing importance with which the large oil companies viewed the acquisition of new oilfields was given to Vicente Lecuna and H. Pérez Dupuy by A.C. Bedford, President of Exxon, at the Second Pan American Financial Conference in Washington. Bedford, who formed part of the same committee as Lecuna and Pérez Dupuy, drew attention to the 'immense value of oil'[70] in the world. However, the American companies still lacked the confidence to invest in Latin America. W.G. McAdoo, the U.S. Treasury Secretary, pointed out in a speech given at the Pan American Society in honour of the delegates attending the conference that U.S. capitalists failed to invest in Latin America because of a dearth of suitable possibilities, but more importantly because they considered local governments were unwilling to protect their investments. McAdoo also stressed that 'nothing would stimulate American investment more in these countries than the security, and real proof when the case arises that such rights will be protected'.[71] When this speech reached Gómez he hurriedly advised Torres to suspend 'all oil concessions until the subject has been better studied, and the amendments to our laws are made during the next session of Congress'.[72] It was also becoming clear that there was an urgent need to draw up a law that would control the industry, and also give people with less powerful connections a chance to participate in the country's new opportunities. The new Decreto Reglamentario for oil enacted on 17 March 1920 sought to curb these inconsistencies[73] but, with corruption on the increase, it did not do so.

The drafting of the first oil law continued under the direction of Arcaya and José Amado Mejía, legal counsellor at the Ministry of Development. On his return to Caracas Lecuna stressed the need to increase government revenues from oil production,[74] and convinced Román Cárdenas, the Minister of Finance (who according to Preston McGoodwin, the American Minister, was the most influential member of the Cabinet), of the need to reduce exploration blocks from 15,000 hectares to 400 hectares, and to increase the government's exploitation royalty to 15 per cent. As Gómez's approval was needed for all this, Lecuna went to see him in Maracay to explain his reasons for increasing the royalty payments. According to Lecuna, companies in the U.S. operating on federal reserve land paid an operating royalty of 15.5 per cent on the commercial value of oil on much smaller concessions than in Venezuela. It was thus reasonable to charge the companies working in the country the same royalty.[75] Lecuna also wanted to have the Delta Amacuro Territory reserved for government exploitation only for reasons of national defence. Preston McGoodwin was horrified upon learning of these proposals, and immediately went to Maracay to confer with Gómez, stressing the point that a reduction in exploration blocks would curtail and diminish the attractiveness which Venezuela held for American capital. McGoodwin reported to Secretary of State Hughes that in the interest of American capital he urged Gómez 'not to permit the project to be considered at the session of Congress which convenes on the 19th'.[76] Gómez answered the American Minister on the afternoon of 4 April to the effect that Venezuela welcomed the introduction of American capital, and that he would do all in his power to 'facilitate all enterprise in which it may become concerned',[77] and that he would not 'countenance any additional changes in the Petroleum Law and expressed almost violent opposition to the plan to restrict areas for petroleum and other materials'.[78]

Notwithstanding this assurance, Lecuna presented a new bill to Congress with all the suggested modifications in early May. According to Lecuna the bill was approved and even printed, but was never published because several influential concessionaires convinced Gómez of the need to modify the terms.[79] Although McGoodwin's representations did have some influence on Gómez, it was nevertheless Julio F. Méndez's acquisition of the concessions (granted to Ramón Ignacio Méndez Llamozas, Federico Wulff, and Lorenzo González Villasmil) which were situated in the Delta Amacuro Territory zone that modified

Gómez's position. Moreover, Méndez had reached agreement
with the Sira brothers and with Márquez Bustillos to take over
their proposals for oil concessions in the same region. In
addition Méndez's attorney, Félix A. Guerrero, had acquired
Rafael Cabrera Malo's concession of 1917 in the same region.[80]
All these concessions were paralysed due to Lecuna's proposal
to declare the region a strategic zone. Torres disagreed with
Lecuna because 'to reserve a zone in that place is the same as in
any other as in the case of a conflagration we will always be a
small country, with or without strategic zones'.[81] The delay was
financially deleterious to Méndez, who wanted Gómez to
approve the concessions. As a result Lecuna's oil bill was
delayed by some of his Senate colleagues[82] so as to allow the
government to present a modified bill to Congress, which was
finally approved on 20 May and published in the Official
Gazette on 29 June.

The law incorporated many of the recommendations made by
Arcaya, Lecuna and Gómez. Exploration concessions were
reduced to 10,000 hectares, with only six such concessions being
awarded to any one person. The exploration period was two
years, at the end of which half the land reverted back to the
government as a National Reserve. The National Reserves were
also extended to cover lakes, rivers, seas and public and
communal lands and were at the disposal of the President to
decide how they should be exploited. Exploitation contracts
would be for 30 years and would be in blocks of 200 hectares (to
be demarcated in chequered form adjoining the National
Reserves) which would be subject to forfeiture if they were not
in production within three years. Taxes were also increased;
now an initial Bs.0.50 per hectare was charged when an exploration contract was taken out, and each exploitation block was
subject to an initial payment of Bs.1,000 per block held; surface
taxes were set at Bs.5 per hectare per annum, and a minimum
exploitation tax of Bs.1,000 per annum levied. In addition a 15
per cent production royalty was imposed to be collected in cash
or kind. This would be reduced to 10 per cent, but never below
Bs.2 per ton of oil, if the oil was more than ten kilometres away
from the port of embarkation or from any other well. The law
also established that only one-fifth of the National Reserves
created each year would be placed on the market annually for

companies to bid on. These National Reserves were subject to the same taxation as ordinary concessions, with the addition that the production royalty was set at maximum 25 per cent of the commercial value of oil. The law also gave the landowners the opportunity to acquire the oil rights to their property.

According to McGoodwin (later repeated by Lieuwen), this law was an experiment to obtain higher taxes and better conditions in exchange for the concessions,[83] but this interpretation is open to doubt. As we have seen, much effort had been expended by government officials to ensure that the law was competitive with that of other countries, while remaining acceptable to the companies. Moreover, one of its main aims was to give the oil companies greater operational security. Therefore, if the law was perceived to be a mere experiment, this objective would not be achieved, and the credibility of the government would be seriously undermined. More importantly, it would have delayed oil development further, and hence government revenues would have remained low from this sector. It can be concluded that the law was enacted for permanent implementation.

The law was a triumph for all those who wanted to see the government increase its share of the country's wealth. Lecuna's 15 per cent royalty was retained, and a compromise was reached over the establishment of the Amacuro Delta Territory as a strategic zone, for article 45 gave the government the right to declare the region a National Reserve when it felt so compelled. Arcaya, too, was pleased about increased taxation. He informed Gómez that if as many concessions were granted during the forthcoming year as in the previous year, then 'revenue to the Treasury would not decrease from the six million bolívars against the several hundred thousands which entered as a result of the concessions recently approved'.[84] This figure was well above the Ministry of Development's total revenue for 1919–20 of Bs.5,058,818, and almost double the total for 1918–19 of Bs.3,799,017.[85]

Property owners immediately took the opportunity to acquire oil concessions over their land. There was an unprecedented increase in the number of concessions awarded during the next year, from 181 concessions in 1920 to 2,374 in 1921, a rise of 1,212 per cent. Between 1922 and 1935 an average of 323 concessions per annum were awarded. Over 55 per cent of the concessions awarded in 1920–1 were concentrated in five states, namely, Anzoátegui (10 per cent), Apure (10.4 per cent), Bolívar (17.6 per cent), Guárico (8.9 per cent), and Zulia (8.8 per cent).[86] This remarkable increase in the number of concessions awarded coincided with the growing interest which foreign companies were taking in Venezuela.

PRODUCTION DISPUTES

With the increasing evidence that the country was destined to become a major oil producer, the government continued its efforts to force CDC either to increase its production or to relinquish part of its massive concession in Zulia. The position for the company had become more difficult because there were now American companies actively engaged in seeking to get the government to annul CDC's concession, and those of other British companies, in order to develop them themselves. As described previously, Torres had in 1918 requested CDC to increase its production. Proctor had agreed to put this request to the Board of Directors in London. However, a year and a half later, no reply had been received by Torres who, rightly, felt that it was 'too long to obtain an answer'.[87] The delay was caused by negotiations which had taken place between Shell and the Carib Syndicate (an American company which in 1919 had acquired Vigas' vendor's rights and 25 per cent of CDC's equity)[88] for the latter company to acquire the remaining 75,000 shares in CDC. Shell was particularly keen to get rid of this concession because the vendor's rights entitled the Carib Syndicate to acquire automatically a quarter of any increase in the company's capitalization, without any cost.[89] These negotiations, however, were unsuccessful, and kept secret from the Venezuelan government, which in October 1919 was forced to take other measures 'as it was the company's clear intention to avoid consideration of the matter'.[90] Torres, therefore, wrote to the company's representative at Caracas, Major Stephen H. Foot, threatening to rescind the contract. The note, despite its serious nature, had a conciliatory tone, as the Minister expressed his 'readiness to discuss the matter in a friendly way'.[91] However, no agreement was reached, and on 5 January 1920 the company was informed that the government was taking it to court.[92] In the ensuing correspondence both parties reiterated their divergent views. Nevertheless, in a final note, Torres was prepared to postpone the court action by a month in order to explore an amicable solution as long as the company recognized that it was at fault.[93] After a great deal of diplomatic assistance from Cecil Dormer, the British Minister, and on account of the government's own anxiety to reach an early settlement, a postponement of the court action was agreed by Villegas Pulido, the Attorney-General, to allow both parties to reach a settlement.[94]

Although the government had shown a willingness to reach a solution with CDC, Dormer believed that it was backed by some of the concession hunters (especially Exxon) who were flocking to the country. Nevertheless, Dormer sympathized with the government's view that the 'concession in question is unduly large, and is a hindrance to development'.[95] Certainly Gómez shared this opinion since, in an interview in early April with Preston McGoodwin, he expressed the desire to press the company to show 'cause why concession should not be annulled for non-compliance with terms'.[96] Consequently, on 7 April he took the matter into his own hands and according to Lieuwen, 'ordered Shell to pay the Bs.3,800,000 annual tax (retroactive to 1915) or give up the concession'.[97] But the government was still looking for a solution, and Villegas Pulido, after conferring with Major Foot, agreed to ask the government for a further postponement of the impending court action. An initial month was granted on 23 April, and extended a further two weeks until 10 June. During this period Major Foot had suggested to the company's head office a possible way out of the impasse. This called for the company to pay Bs.50,000 annually in exploitation tax which would be 'amortised when exploitation of oil actually begins'.[98] The London office deemed this suggestion inadvisable since the company had decided to 'stand on our absolute rights and not to accept arrangement suggested'.[99]

At this juncture the government, encouraged by Exxon, 'who may be making determined efforts to turn out the British Companies'[100] in order to acquire the concession themselves, was also considering disputing the concessions held by VOC and the British Controlled Oilfields Ltd. The companies and representatives of the Foreign Office had urgent talks with Pedro

César Domínici, the Venezuelan Minister in London, to see whether matters might be explained to Gómez directly. On 11 May Domínici sent a telegram to Gómez in which he described the intended action by Torres towards the companies as causing 'deep dismay among financial circles interested in oil developments'.[101] He added that 'in the oil interests of Venezuela I beg you await mail dispatched this day before continuing proceedings'.[102] The following day a long letter was sent to Gómez in which he set out the adverse effects which the continued litigation would have on the attraction to British capitalists of investing in the country. He also explained to Gómez that Torres wanted the companies to pay Bs.2 per hectare surface tax on the entire concessions which they held. This would mean that the British Controlled Oilfields Ltd, for example, would have to pay an additional Bs.1 million per annum in tax (other companies slightly less), and this 'according to the interested parties amounts to the abandonment of their oilfields'.[103] Thus the effect of disputing the companies' contracts would be to deter foreign investors from entering the country, and to further delay production by the companies already in the country. Duncan Elliott Alves, the Chairman of VOC, and of British Controlled Oilfields Ltd, also warned Gómez that any court action brought against the British companies 'would be disastrous for the good creditworthiness which Venezuela enjoys thanks to your efforts',[104] and that this would have detrimental effects 'even worse than we can imagine'.[105] The Carib Syndicate also viewed with great anxiety the events which were occurring in Caracas, because its investment in the company, valued at $10 million, would be lost with the rescinding of CDC's concession by the government. C.F. McFadden, the Director of the Carib Syndicate, officially requested the State Department to assist CDC in its fight against the Venezuelan government.[106] But the U.S. government had no intention whatsoever in assisting a British-controlled company out of its troubles in Venezuela.[107]

After a week there was no significant change, so the Foreign Office, seriously worried, sent a telegram to Dormer on 18 May indicating that he should take 'all possible action to prevent any reduction in areas for which concessions have been obtained'.[108] The indications were that the government was acting at the direction of Juancho Gómez, and that Gómez was afraid to interfere.[109] Dormer considered these to be exceptional circumstances, and as such warranted diplomatic intervention, despite the fact that there were signs that the government were 'looking for a way out of the crisis, by a solution that will hurt no one'.[110] The government, however, seemed to have no clear way out of the imbroglio. On 7 June Dormer saw Gil Borges, the Foreign Minister, who said that he had not received any information from Domínici about the grave concern with which the British oil companies viewed Torres' impending court action. Dormer was surprised because he was under the impression that Domínici had informed Gómez of the situation in London.[111] Gil Borges was well aware of this, as he commented to Torres on 14 July that Domínici 'seems more like a Colon employee than Venezuela's official representative'.[112] Gil Borges was also aware of the possibility of U.S. intervention on behalf of the American quarter-interest in CDC. He had received the June despatch from Santos Domínici (Pedro César's brother and Venezuelan Minister in Washington) in which he was informed, albeit two months late, that Julian Arroyo, acting on MacFadden's behalf, had been in Washington to gather assistance for the protection of the company's stake in CDC.[113] Gil Borges' reply to Dormer was therefore a delaying tactic to allow a settlement in which the government would not lose face to be reached. The arguments expounded by Domínici and Alves were potent enough to convince the government of the need to reach an amicable solution. The loss of confidence in Venezuela among the international banks and capitalists of London willing to invest in the country would not only have meant that an important source of credit would dry up, but would also make the country wholly dependent on American capital which, in the light of the cool relations between the U.S. and Venezuela since 1917, was unacceptable. Furthermore, this dependence on the U.S. would have far-reaching effects, allowing American oil companies a free hand in the development of the country's oil industry. Gómez wanted to avoid this at all costs as he felt that it could lead to a possible source of friction in the future, and eventually to U.S. intervention.

Although the State Department had refused up to then to assist the American interests in CDC, the Venezuelan government knew that the possibility existed of the U.S. entering the dispute at any time. Dormer was also sure that Gómez did not want to alienate British capital in Venezuela, and was convinced that once this was made clear to Gómez the whole débâcle would be quickly wound up. He therefore decided to address a Note to Gil Borges in such strong terms that he would have no alternative but to pass it on to Gómez. In it he stated that the British government 'does not recognise any reduction in the area of the concessions acquired by legal contract between the government of Venezuela and the British companies, unless such a reduction is freely agreed to by both parties'.[114] Further: 'Any measure which weakened British interests would be at complete variance with the assurances received from General Gómez, and the Provincial President, that such interests would be protected, thus encouraging further British capital to these lands.'[115] He ended by saying that he had written the Note because he was convinced that the government 'is unaware of the potential danger of adopting such an attitude and policy, and feel that by addressing Your Excellency in this friendly manner, you will take into account my reasons for preventing a possible disagreement between our two countries'.[116] Two days later Gil Borges forwarded Dormer's Note to Gómez as requested by the British Minister.

The Cabinet nevertheless decided unanimously in early June that the company had been given sufficient time to reach a settlement, and that proceedings to have the concession annulled should begin. On 12 June, two days after the agreement between the government and CDC to postpone court action had ended, Villegas Pulido entered a suit at the Federal Court of Cassation requesting the court to annul CDC's concession (though allowing it to retain the three blocks it had previously denounced) or to force it to pay the taxes demanded by the government. Similar action was to follow against VOC and the British Controlled Oilfields Ltd.[117] If the court's decision was unfavourable to the company the long-term effects on British oil companies would be momentous, as the decision implied that 'Shell's Aranguren concession, British Controlled's Planas concession, and North Venezuelan Petroleum's Arraiz concession were subject to the same treatment.'[118]

At this time six of the largest American oil companies were in

Venezuela, and McGoodwin reported on 11 June that they were 'confident of the ability to secure contracts covering these properties'.[119] Moreover, as we have seen, McKay, acting on behalf of the Sun Oil Co. of Philadelphia, and Exxon offered in April to pay the Ministry of Development $1,350,000, and an equal sum to Gómez, for the privilege of exploring for one year the properties of CDC, VOC, and the British Controlled Oilfields Ltd.[120] According to McGoodwin the prospects of acceptance appeared to be 'very favourable'.[121] It was no wonder, then, that Dormer felt that 'American secret support of Government's attitude is more patent than concealed'.[122] The State Department's aim was without doubt to have CDC's concession annulled,[123] but the evidence suggests that they did not actively intrigue to achieve this end. They simply refused to render any diplomatic assistance to companies which were 'either British-controlled or closely affiliated with British-controlled companies',[124] which included *inter alia* CDC and the Carib Syndicate.

CDC at this juncture appeared to be in a hopeless position. Villegas Pulido, at a social function in the evening of the very day on which the suit had been entered at court, informed McGoodwin that the annulment of the company's concession was just a matter of time for publication in the Official Gazette. The large American oil companies referred to by McGoodwin were confident of securing the properties in question, and the 'approval of same prior to adjournment of Congress'[125] on 27 June. Despite Villegas Pulido's clear assurances, no announcement appeared in the Official Gazette. McKay attributed this to the delay in receiving from the companies he represented definite information on the bids submitted for CDC's property. However, legal complications had arisen. As was the usual practice, the court's decision had been prepared by the one member of the court to whom the case had been referred. This decision was not binding until the entire court had voted for it. On 23 June Gil Borges informed McGoodwin that this meant that 'confirmatory steps'[126] would have to be taken by the court *en banc* before a decision could be reached.

During this time the State Department was in the process of modifying its policy in the case of Vigas' vendor's rights. Vigas, the original concessionaire, had included a clause in the transfer deed to CDC which stated that if the company did not develop the property to its full capacity then the concession reverted back to him. As a result, in a final attempt to move the State Department to assist the Carib Syndicate, MacFadden wrote to them on 16 June stating that the annulment of CDC's contract would also destroy the vendor's rights held by the company. However, if the State Department were to persuade the Venezuelan courts to respect these rights then the Carib Syndicate would 'succeed to all rights of the original holder of the concession'.[127] The State Department saw fit to support this claim, and cabled McGoodwin on 24 June requesting that the government of Venezuela be informed that the U.S. government would be pleased to see the Carib Syndicate's equitable rights in CDC 'recognised and protected'[128] by the government, which McGoodwin did three days later. In the meantime CDC's lawyers introduced a motion that the court did not have the legal power to decide this matter. This was rejected on 16 July, and the original suit continued.

Although there were 'strong influences at work',[129] Torres was still looking for a way out. Both legal advisers whom Torres consulted concurred with the view that CDC had the legal right to keep its concession unexploited so long as it had paid the minimum surface and exploitation taxes.[130] It was, therefore, clear that although the company had the legal right to retain its full concession, it would have to develop it fully or pay the minimum surface and exploitation taxes over the whole concession, which amounted to a tax bill of Bs.3,800,000 annually. The company was unable to pay such a heavy burden, especially on a hitherto unproductive property. Nevertheless, in August, in order to end the dispute, the company proposed to pay Bs.40,000 annually as a minimum exploitation tax. This proposal was rejected by Torres,[131] and a stalemate ensued, dragging on for several more months. Once again, in October 1920, in another visit to Gil Borges, Dormer told him that the dispute had gone on for a year, and again threatened the Venezuelan government by declaring that if a satisfactory solution was not found the British government would have to 'reconsider their general attitude'.[132] But by this time the storm had passed as there was 'no great danger at present of our oil interests in Venezuela being injured'.[133]

As the months went by the prospects for an amicable arrangement between CDC and the government had improved. Shortly after the negotiations between the company's representative and Torres had broken down, Santiago Vega, an 'intimate friend of the Provisional President',[134] informed Dormer that the Attorney-General had been in Maracay and was of the opinion that, legally, CDC was right, and that unless Torres 'showed himself more conciliatory he would lose his place'[135] in the government. This threat, together with Gómez's direct intervention, made Torres change his attitude. On 12 February 1921 a settlement was reached whereby the company would retain its full concession. The company, however, agreed that during the next five years it would select blocks for further exploitation. At the end of this first five-year period the company would pay an additional surface tax of Bs.0.20 per hectare per annum on the land selected. A second five-year period would follow during which the company would definitely decide which tracts of land it would exploit (the land which the company did not select would revert back to the government). Surface and exploitation taxes would remain at Bs.1 per hectare per annum and Bs.2 per ton respectively. On 16 February 1921 the Cabinet ratified this settlement with CDC,[136] and on 25 April of the same year a similar settlement was made with VOC.[137] The outcome of this affair was that the British had been successful in repelling the American threat. In 1922 Shell further strengthened the legal position of the Valladares concession by paying the government Bs.10 million to have its titles reconfirmed.[138]

The agreement reached by the British oil companies was a considerable achievement since the government could have easily rescinded the concessions and have awarded them to other companies present in the country at the time. The government would certainly have benefited in the short run for the concessions would then have come under the much more onerous tax provisions of the 1920 Oil Law. These short-term benefits were, however, rejected in order to establish a sound basis on which companies of various nationalities could flourish. By rescinding the companies' concessions Gómez would certainly have created an international problem for his country (resulting in a stalemate similar to that in the Middle East at the time), with Venezuela, and consequently Gómez's government, being the real losers in the fracas. Gómez, wanting to avoid foreign intervention, did not care to see the industry developed exclusively by American oil interests, for that would have placed his government in a very dependent position, both in political and in economic terms. By retaining the British oil companies he not only avoided direct confrontation with a British government which was actively seeking to lessen its dependence on American oil sources, but he could also continue afterwards to play one nationality against the other, thus achieving greater revenues, and a certain amount of independence and control over the development of the industry. The agreements can thus be seen as a long-term victory for the government. By agreeing to limit the extent of their concessions after an initial exploration period, the companies were forced to relinquish their claim to be the sole owners of the whole of their concessions. The British companies would have to develop their concession to the fullest extent or lose the valuable areas to their rivals. Now that Shell definitely had to develop its property, its main rival, Exxon in order to remain competitive, would of necessity have to enter the country in a significant way.

INTERMEDIARY OIL NEGOTIATORS

The arrival of Exxon and other large American companies at this juncture meant that many of the concessionaires could transfer their concessions to these companies. Gómez's policy of flooding the American market with concessions would then have the predicted effect of attracting more companies to the country, now that the two largest oil companies had made their decision.

Some of the new concessionaires had the expertise and connections to negotiate their respective concessions directly with the companies, but the vast majority of them did not possess these qualities, thus allowing a few key intermediaries, in many instances closely connected to Gómez or to his entourage, to negotiate individual concessions, or a mixed package deal, with the companies. For example, Bernardino Mosquera and José María García Gómez transferred their concessions to the British Equatorial Oil Co.,[139] whereas the same company acquired the concessions belonging to Pedro

Vicente Navarro and José M. Capriles through Howard R. Stewart. The same José María García Gómez had to entrust José A. Domínguez, who had formed the Venezuelan Petroleum Syndicate to deal in oil concessions, to sell his concession in New York.[140] For those lucky enough to obtain concessions in known oil-bearing lands, the transfer price was very high: José María García, for instance, transferred his coastal zone concession in 1927 to the Maritime Oil Corp for $500,000 and a 5 per cent production royalty.[141]

Although the acquisition by property owners of the subsoil oil rights to their property had diminished the areas in which concessions could be awarded, there were still extensive zones, such as lakes, rivers, coasts, and *tierras baldías* (uncultivated common land), where the government had the discretionary right to award concessions (which had to comply with the obligations set out by the 1920 Oil Law and which needed congressional approval), with the result that there was still considerable scope for concessions to be awarded to Gómez's close relatives. In March 1921 Juancho Gómez requested Torres to award a concession to Colonel Elio Rivas Rojas.[142] In December 1920 Ignacio Andrade (another of Gómez's sons-in-law) and José María Márquez (son of Victoriano Márquez Bustillos) solicited oil concessions from the Ministry of Development.[143] From the available evidence, it was José Vicente Gómez who was most active in this sphere. In July 1920 José Vicente recommended to Torres 'very specially'[144] his brother-in-law, Carlos Delfino, and José M. Capriles in order that an oil concession in the Miranda District of Falcón State[145] might be awarded them. In December of the same year he requested Torres to provide Delfino with the information on all the concessions on Lake Maracaibo in which he was interested.[146] José Vicente later sent the names of the persons purporting to have been awarded them. Torres appears to have objected to this procedure but, after receiving Gómez's approval,[147] agreed, that 115 contracts of 10,000 hectares each, and one of 9,000 hectares, should be awarded to the following people: José Andrade, R. Aspurúa Feo, Antonio Báez, A. Brandt C., R.M. Clemente R., G. Franco Golding, B. Hernández C., José Izquierdo, Alejandro Jiménez, J.M. Landa, F. Castro López, J.B. Madriz, C. Márquez M., P.J. Morales, R.R. Revenga, F. Reyna H., A.J. Plaza, O., Celso Serha,

Baldomero Uzcátegui García, and José Valero Lara.[148] Except for Baldomero Uzcátegui García, who received two concessions, all acquired six concessions of 10,000 hectares each. On 13 June 1921 the contracts were approved by Congress, and appeared in an extraordinary issue of the Official Gazette on 16 August 1922. This was not the only venture in which José Vicente and Carlos Delfino co-operated. Together they held 37 other oil contracts (16 in Sucre State and 21 in Yaracuy State), which in 1925 were transferred to the Central Venezuelan Oil Corp.[149] José Vicente continued recommending people to Torres for him to award them concessions: in May 1921 his brother-in-law José Manuel Revenga, was recommended,[150] and Torres was also requested that the concessions which had previously belonged to the Val de Travers Asphalt Paving Co. Ltd be awarded to Pedro Hernández G.[151] An early success was achieved when Delfino transferred several concessions to the Sun Oil Co. The partnership even wanted to deal in the National Reserves created by the 1920 Oil Law. Torres, however, firmly rejected this idea, explaining that this could only occur after 1922 when exploration had ended, because only then would the true value of the National Reserves be known. If they were granted now, the government would be selling a rich deposit at a derisory sum.

Other important intermediaries outside the immediate Gómez family were the Capriles brothers.[152] In 1921, for example, Abraham Capriles transferred to S. McGill, who represented an American company, Lorenzo Carvallo's coastal concession in Sucre State for $420,000. Abraham considered the deal to be good, 'given the fact that there is a real avalanche of sellers, and last year's experience when the price dropped to Bs.2 per hectare could be repeated'.[153] Moreover, the deal had the added attraction of drawing the company's attention to Raúl Capriles' own concession of 140,000 hectares in Sucre State. Antonio Aranguren also acted as an intermediary for the Capriles brothers, and for Arcaya and Mellado, assuring them before he left for New York in August 1921 that he could place their respective concessions with various American companies at $4 per hectare.[154] But the companies were becoming impatient at dealing with intermediaries and Aranguren returned to Caracas in October empty-handed, informing Abraham Capriles that the oilmen 'do not want to enter into contracts with intermediaries, but only with concessionaires or with persons properly authorised or with power of attorney'.[155] Others were even more blunt in their approach: for example, M.V. González Rincones proposed to Urdaneta Maya, Gómez's Secretary-General, that both should enter the oil business to make 'a million dollars in a couple of years'.[156] González, whose friends included the directors of the Barco Petroleum Corp. (developing the Barco concession in Colombia), wanted Urdaneta Maya to secure concessions for him in Zulia, Coro Gulf, and on the Coro Gulf island of Maragua, for the construction of a refinery or an oil storage depot.[157]

OPPOSITION TO THE 1920 OIL LAW

As we have seen, the oil development of the country was proceeding well. The importance of the country to the international oil fraternity can be gauged by the American Petroleum Institute's invitation to Torres to attend its annual conference to be held in New York in November 1920. In preparation for the large-scale development of the country's resources, Torres considered the creation of a Dirección de Petróleo within the Ministry of Development to look after the oil interests of the country exclusively, and a Dirección General de Administración to collect more efficiently the large prospective oil revenues.[158]

However, soon after the 1920 Oil Law was enacted, there were ominous signs that the companies would not accept the law as it stood. According to McGoodwin, the representatives of three American oil companies in Venezuela had 'instructions from their companies to remain here and watch developments and to render this legation any assistance in the nature of informal representations to the Minister of Fomento [*sic*], the Counsellor of the Minister of Fomento [*sic*], the Provisional President, and other officials of the Government'.[159] Several meetings were held at the U.S. Legation to determine which articles of the law needed amendment. In March 1921 representatives of other American companies which wanted to acquire exploration titles joined the sessions and submitted memoranda to each other 'in which they set forth their respective ideas as to the changes which should be made in order to facilitate the industry invariably with due regard to the interests of the Government of Venezuela'.[160] The companies then submitted a joint Memorandum to the Ministry of Development detailing their objections to the present law. They required a new law which would permit them to hold as much land as they could pay taxes on, instead of the 60,000 hectares allowed by the present law (article 7); property owners should not have the indefinite subsoil oil rights (article 8); exploitation blocks should be 1,000 hectares, instead of the 200 hectares allowed by law, and not in chequered form (article 31), and the original concessionaire should have first choice in bidding for the National Reserves created on his land; the exploitation time limit of 30 years (article 32) should be extended until the companies could produce oil in commercial quantities. The companies also found that the surface tax of Bs.5 per hectare per annum on each exploitation block of 200 hectares was excessive. A more acceptable figure would be Bs.0.05 per hectare during the exploratory period, increasing to Bs.1 during the first two years of exploitation and with further increments of Bs.1 every two years, reaching a maximum of Bs.5 per hectare per annum. However, the operator should not be under the obligation to pay both the surface and production taxes, and therefore as the latter was expected to greatly exceed the former, it should be applied as credit against surface taxes.[161] In addition, the production royalty on both concessions (15 per cent) and National Reserves (25 per cent) was too high; a more acceptable figure would be Bs.2 per ton of oil produced or 10 per cent of the commercial value of the oil, whichever was the highest. The government's intention of releasing for competitive bidding only 20 per cent of the annual National Reserves created was also too small. It was felt, too, that the provision of article 50, which stated that exploitation blocks must be in production within three years, should be eliminated because 'sufficient stimulus for development is provided for by the surface tax'.[162] The companies finally sought more generous exoneration from customs duties on supplementary materials and equipment for road construction, warehouses, hospitals, etc.

These objections must have come as a great surprise to Torres who had spent many long hours, and much effort, studying the oil laws of other countries. Still, the government's immediate objective was to establish a strong and secure oil industry, and so it was flexible enough to study the proposals in order to reach a compromise with the companies. In any case the law would have to be changed in 1921 because the land-owners' right to the subsoil on their property would come to an end, and the law would therefore be invalid because it contained provisions which would no longer be applicable.[163] This would give the government the opportunity to correct the minor inconveniences which had arisen during the previous year in order to attract more investments. Torres agreed that article 50 of the law had to be changed because the cost to the companies would be enormous. As the law stood, the maximum number of exploitation contracts awarded to any one operator would be 150, which meant that drilling equipment would have to be used in each block to comply with the law that demanded that each block should be in production within three years. A minimum outlay of between Bs.153,000 and Bs.683,000 per block would be incurred, in addition to the Bs.1,454,000 which was estimated would be spent in storage facilities, access roads, pipelines, etc. Torres reckoned that in the next few years a minimum of 7,500 exploitation blocks would be awarded, necessitating the importation of 1,250 drilling rigs.[164] Thus he agreed to drop this provision of the law as it served no good purpose, and proposed instead that the exploitation blocks would be annulled if the operator failed to pay the annual surface taxes. The government would then be assured of a minimum annual revenue, and it would serve as a stimulus to production.[165] However, Torres was not so pliable with regard to the operators' request to group together the five exploitation blocks into one large block. He argued that the country would not benefit from such a change. But the fear that the oil companies might pull out of the country altogether spurred on the large number of oil concessionaires (led by the *gomecista* group) to place the greatest pressure on Torres to change the 200-hectare provision.[166] Márquez Bustillos (secretly a concession-holder himself) lobbied Gómez to approve the change in the new law that was being drafted, and sent Urdaneta Maya, Gómez's Secretary-General, a Memorandum in which he argued that if the blocks were not allowed to be grouped together then both the country and the concessionaires would lose because the companies would not invest.[167] The fear that one company would monopolize production (the main reason for establishing the 200-hectare exploitation block) was valid only in the event of oil being found, as it was logical to assume that oil would be discovered in the adjoining blocks. But in Venezuela's case, production started only after the blocks had been demarcated. Thus, by obtaining a 1,000-hectare block, the operator had the same probabilities of finding oil as in the five blocks adjoining the government's blocks. Finding oil would only reinforce the view that the region as a whole was oil-bearing.[168] A further objection to the 1,000-hectare block was that it was much bigger than similar blocks awarded in other oil countries. The available evidence did not substantiate this claim. In the U.S., where the subsoil rights were vested in the landowners, the companies held concessions which were as great as the land occupied, and on Federal lands the oil operator could take up one-third of the land in one block. The same applied in Mexico, prior to the 1917 Constitution, and in Colombia, where exploitation blocks were 200 hectares with no restrictions on the number held. The Memorandum concluded that the whole question should be examined from a practical point of view which would consider the effect such a provision would have on the future development of the industry. If it was found that the companies were unwilling to invest because they could not group together five exploitation blocks, then it was obvious that it should be allowed. As all the signs pointed to the companies not investing unless this provision was changed, and as Torres was adamant that the provision would remain, Márquez Bustillos proposed to Urdaneta Maya that Congress should add a clause allowing five blocks to be grouped together.[169]

In the meantime, McGoodwin went to San Juan de los Morros where he spent six days with Gómez during which he 'discussed the necessary changes'[170] to the law. Gómez assured him that these would be made to 'suit the requirements of the petroleum development companies during the present session of Congress'.[171] Torres had already presented his new oil bill with three minor modifications to Congress when Juancho Gómez made 'informal representations'[172] to the members of the Permanent Development Committees of the Chambers of Deputies and Senators. As a result, on 25 May, two additional bills were submitted to Congress, and the following day the head of the Senate Development Committee called at the U.S. Legation to inform McGoodwin that the 'two Committees had decided to draft, jointly, a Bill which he said "would include various other reforms"'.[173] That same afternoon Márquez Bustillos informed McGoodwin that a 'new Bill would be presented to the Congress which would meet the features of the present law, which all of us have found to be so objectionable'.[174] The new proposal which Arcaya and McKay helped to draft was sanctioned by Congress on 16 June 1921 and became law on 11 July 1921.

The new law increased the maximum exploration and exploitation contracts allowed to 240,000 hectares and 120,000 hectares respectively. Exploitation contracts would be for 30 years, and operators would now be able to group together five 200-hectare exploitation blocks to form one large block. The obligation to put into production these exploitation blocks within three years was also dropped, with the companies only having to pay minimum taxes to retain their rights to them. Taxes were an initial Bs.0.50 per hectare for an exploration contract, and Bs.5 per hectare payable within a month for an exploitation contract. Annual surface taxes were set at Bs.5 per hectare, and production taxes would be 15 per cent payable in cash or kind (or 10 per cent if the rig was 40 kilometres away from the nearest loading quay or more than 10 kilometres away from another well). A 25 per cent production royalty was still retained in the case of National Reserves.

THE 1921 OIL LAW IS ALSO UNACCEPTABLE

Despite the fact that many of the oil companies' suggestions had been incorporated into the new law, it soon became apparent that the operators were not satisfied with it. It was severely flawed as a result of poor drafting, and in parts even contradicted itself. Fred H. Kay of the Venezuelan Sun Ltd submitted a Memorandum to Torres in order to clear up some of the ambiguous articles (articles 32, 41, 73 and 74) in the newly enacted law.[175] Goethe of the Pure Oil Co. and Louis Scholl of The Texas Corp. also deemed the law ambiguous in many important points because it had been drafted 'less with a view to the interests of practical oil operators than to the native concession holders, who naturally hope to derive the maximum profit for their leases from foreign purchasers'.[176] A more serious criticism, now that oil prices had dropped in the U.S., was that the taxation base was too onerous. As a result, the American companies had not acquired any of the 2,300 concessions awarded between July and October, and the concessionaires had once again started to place pressure on Torres to get him to reduce taxation.[177] However, he considered lower taxation to be counterproductive for the country for a number of reasons. First, the Shell group of companies, the largest producers, would be able to adapt their already liberal contracts to the new law, thus reducing oil revenues to the Treasury from this source. Second, the large number of contracts had not been transferred precisely because there were too many of these. It was unlikely, Torres argued, that more than 50 per cent of them had any commercial value and would be transferred to oil companies because oil would not be found over the entire country. A reduction in taxes would not necessarily herald a rush to take up more concessions, but rather would mean that future oil revenues would be reduced to an 'insignificant amount'.[178] Third, a comparison of oil legislation in other countries would reveal that Venezuela's taxes were not onerous. Nevertheless, over the next few months, it became apparent that the American oil operators and the oil concessionaires would 'make a concerted effort to secure more favourable legislation in the spring'.[179]

Upon more detailed study of the law Torres concluded that there was room for improvement, not in order to reduce taxes but to rectify some of the flaws and contradictions found in it, such as articles 32 and 53 which effectively annulled surface taxes.[180] Torres also suggested that rather than grant the companies an extension to their exploration deadlines as requested, the concession should instead be considered to be under exploitation (with half of it becoming part of the National Reserves) in order to force the holder of the concession either to pay the minimum annual taxes or to renounce the concession.[181] In order to forestall any appeals to Gómez to reduce taxes, Torres mounted an elaborate campaign which started in January 1922 when the Minister's *Boletín* published details of the Mexican oil experience, including the increase in oil export taxes in 1921, the 1921 oil bill, and a sample of an oil contract.[182] In a lengthy letter to Gómez, Torres explained that Mexican tax rates were much higher than Venezuela's, the companies having to pay an exploitation, surface, and export tax of 5 pesos (slightly over Bs.10), whereas in Venezuela 'the annual surface tax . . . is practically void, and the 10 per cent production tax, fair and reasonable, will always be reduced to Bs.2 per ton as long as the government cannot receive it in kind'.[183] In addition, the Mexican government granted oil contracts on the condition that the companies or concessionaires adapted their contracts to any future oil laws, a clause which Torres felt was 'very tough',[184] but nevertheless 'every month contracts are signed, and the Mexican oil industry, contrary to expert prediction, remains active'.[185] Torres thus advised Gómez:

when México, a disorganized country, and in perpetual war, proceeds in this manner with its oil, we must, I feel, consider carefully the contractors' intentions of acquiring here, almost without tax, so many concessions, and also to take into account that your Administration has been in all official activities, and in particular in fiscal matters, the most far-sighted we have had to date.[186]

Although Torres was correct in stating that oil companies continued to invest in Mexico in 1921, the tax increases of the same year[187] did precipitate a crisis in which the companies refused to export oil until a reduction in taxation was achieved. The companies encountered problems elsewhere. In Colombia, for example, the petroleum law of 29 December 1919 was also, in their view, onerous;[188] moreover, any development could only take placce after the signing of a Colombo-American treaty regularizing diplomatic relations between both countries,[189] which was achieved at the end of 1921. In the Middle East, too, American oil companies encountered difficulties. The initial British 'closed door' policy had led to long, protracted, negotiations, culminating in 1928 when American oil interests were allowed to participate in the development of the Turkish Petroleum Co.'s concession in Iraq.[190] Although the concessions awarded in the Middle East were more generous and profitable than those obtained in Venezuela, there were major drawbacks in the early stages of development in that region because of its natural inaccessibility, both in political and in economic terms. Venezuela's geographical position placed her at an advantage to both her Mexican and Middle East competitors. It is doubtful that Torres was aware of all these developments. What is certain is that he saw no reason why the same companies which operated in Mexico, and willingly paid the pre-1921 tax of Bs.10 per ton of oil, should not accept Venezuela's much lower tax levels. Moreover, it was apparent that the operators had no intention of leaving Venezuela. It was his conviction that once production reached Mexican levels, Venezuela would soon cover all her budgetary needs from the minimum exploitation tax of Bs.2 per ton. Nevertheless, Torres commissioned a study to look at the most effective way of taxing oil production. The study looked at two types of taxation, one as a percentage of the commercial value of oil, and the other as a fixed amount (ranging between Bs.2 and Bs.3.50) to be levied on the specific gravity of oil. Under the first system a 12.5 per cent royalty (the best offered by a company) would yield the government a return, under 1922 prices, of Bs.1.94 per ton (Bs.1.06 in 1921), whereas under the second system the government would be able to levy a tax of Bs.2.25 per ton.[191] In the end Torres opted for a system giving an assured minimum rate and which took into account any future increase in the value of oil. In March he submitted his newly drafted bill to the Cabinet, ensuring, as he explained to Gómez, that all was 'clearly expressed in order to avoid all future controversies arising from its interpretation . . . [and feeling that] there is no ambiguity'.[192] He also took the unprecedented step of publishing the draft in the Ministry's *Boletín*[193] to give it the widest possible publicity among the concessionaires. The Cabinet agreed to the provisions set out by Torres, especially those dealing with taxes, and on 24 April Pedro M. Arcaya, President of the Senate, presented Congress with Torres' Proyecto de Ley de Hidrocarburos. The bill's most important modifications to the law concerned taxation and the curtailment of speculation. It called for the following taxes to be levied: for an exploration concession, an initial surface tax of Bs.2.50 per hectare in the first year, and Bs.0.10 per hectare in the subsequent two years; for exploitation contracts, surface taxes would be Bs.3 per hectare per annum for the first three years, and subsequently Bs.4 per hectare per annum, and production taxes would be Bs.2 per ton when the value of oil at Venezuelan ports was Bs.20 or less, increasing by Bs.0.25 for every Bs.10 increase in the value of oil. In order to curtail speculation in oil concessions, and to stimulate production, an operator was to make a deposit for each type of concession awarded as a guarantee of good faith, the money to be returned when it was confirmed that operators had 'established permanent works or construction in their concessions, whose value exceeded ten times the value of the guarantee',[194] or if no oil was discovered.

The bill ran into trouble immediately when it was considered by the Senate Permanent Development Committee, and there is little doubt that influence was brought to bear on the committee to reduce taxes and abolish the monetary guarantees of good faith. Two of its seven members, Dr Luis Felipe Blanco and Dr J.A. Tagliaferro,[195] held oil concessions in their own right,[196] but more importantly, Dr Tagliaferro was Márquez Bustillos' personal secretary. Relations between Torres and Márquez Bustillos were so acrimonious that when the latter omitted to mention the Ministry of Development in his Annual Message to Congress, Torres tendered his resignation. This was retracted the following day when Dr Tagliaferro assured Torres that Márquez Bustillos' error had arisen from a misreading of the text.[197]

The Senate Permanent Development Committee thought that Torres' bill would inhibit production in a large number of concessions. Torres countered by saying that the extensive hectareage under oil concessionaires ensured by default that many concessions would not be transferred. In order to avoid the risk of speculators whose sole interest was dealing in concessions and not developing them, Torres had introduced the money guarantees which, at a time when large oil companies held extensive areas, would ensure that only serious concerns entered Venezuela. Torres also dismissed the notion that the initial exploration surface tax of Bs.2.50 per hectare was prohibitive, as past experience had shown the contrary. For example, in 1919 McKay, Rafael Requena, and Julio F. Méndez had all transferred their concessions in New York for between Bs.10.40 and Bs.20.80; and in 1920 Capriles transferred 75,000 hectares at Bs.3.80 per hectare, José María García Gómez transferred 45,000 hectares at Bs.12.30 per hectare, and Pedro Navarro transferred 75,000 hectares at Bs.6.93 per hectare. Thus by introducing a fairly high initial exploration tax, compared to the previous level, Torres hoped to encourage responsible companies and to avoid awarding contracts to people or companies whose sole object was to sell them to other companies, thereby delaying the exploitation of the oilfields. Furthermore, by eliminating intermediaries the companies would save money under the proposed bill. For example, Warren W. Smith had recently acquired from a concessionaire 75,000 hectares for Bs.290,671.65, of which Bs.286,000 was the transfer fee. Under Torres' scheme the same area would cost Bs.214,837.50 (including legal fees), a saving of Bs.75,834.15, or 26 per cent, on the original price. The scheme also eliminated the delay associated with intermediaries as well as increasing government revenues. It would not be difficult to attract important oil concerns, as critics predicted, because companies would negotiate concessions directly, bypassing the intermediary.[198] Critics also observed that unless companies were given special tax advantages they would not prospect in the remote areas of the country. Torres found this view absurd because the companies could construct pipelines without any real loss of profitability, so that there appeared no logical reason for the government to subsidize companies operating in distant regions.[199] Torres also rejected outright the notion of reducing taxes. A final criticism of his bill was that the oil companies would only concentrate their operations in known oil-bearing lands, while the rest of the country remained outside their scope. Torres interpreted this as being beneficial to the country as a whole, because the National Reserves, which would be coming on the market for the first time later that year, would reach high prices due to the competitive bidding of the companies. Companies would prefer to acquire National Reserves situated in known oil-bearing lands rather than concessions in unknown lands. In addition, it was against the country's interest for all its oil to be exploited at once, and it was also doubtful that oil existed over all of Venezuela.[200] Proof of this, according to Torres, could be seen in the two recent proposals presented to the Ministry to acquire National

Reserves; in one case the anonymous company had offered
Bs.200,000–250,000 to acquire 100,000 hectares in the Perijá
District of Zulia State; and in the other, a company had offered
Bs.35,000 for 150 hectares, also in Zulia, in a zone which had
been explored and found to contain oil. These sums, as Torres
explained, were what any concessionaire would obtain by transferring his concession to a company. By eliminating the intermediary a new source of revenues would be created for the
government.[201]

There were other people who urged Gómez and Torres to
increase taxes rather than decrease them. Vicente Lecuna and
Pérez Dupuy were two who criticized Torres for doing away
with the minimum 10 per cent production royalty in the
proposed oil law. They reasoned that if the price of oil was
Bs.120 per ton, Venezuela's share under the present oil bill
would drop from Bs.12 to Bs.4.50 per ton.[202] Lecuna and Pérez
Dupuy, who had been advised in Washington by 'the foremost
experts in these matters'[203] not to allow the country's production tax to fall below 10 per cent, felt that oil was such a
valuable commodity that when 'the companies realize that the
government will not yield to their pressure then they will
acquire concessions at any price, agreeing to pay tax at 10 per
cent'.[204] Their argument was based on the misconception that
royalty payments were calculated on U.S. oil prices, instead of
on the value of the oil in Venezuela. As there was no local
market for Venezuelan oil, its commercial value was determined
by the costs of production; consequently, the Ministry's calculation had shown that 'two bolívars were equivalent more or
less to 10 per cent in kind, bearing in mind that a ton of oil at
the port of embarkation is worth twenty bolívars'.[205] Moreover,
increased oil production would lower production costs per ton
produced, with the result that under the Lecuna–Pérez Depuy
scheme government revenues would also decrease. This problem was avoided, however, under Torres' scheme by retaining a
minimum production tax, and in addition, if the value of
Venezuelan oil increased above the minimum value of Bs.20 per
ton set by him, then a surtax of Bs.0.25 per Bs.10 increase per
ton would be levied. Under the Lecuna–Pérez Dupuy scheme
the companies would be forced to abandon their operations in
the country because of excessive taxation. For example, the

average U.S. oil price for 1922 was Bs.31.2, which meant that companies would have to pay a production tax of Bs.3.12, much above the Bs.2.25 proposed by Torres.[206]

Nevertheless, those with vested interests among the *gomecista* concessionaires, and intermediaries with easy access to Gómez's ear, convinced Gómez of the need to change the law. Consequently, Dr Rafael Hidalgo Hernández, who had also co-operated in preparing Torres' oil bill, was entrusted with the task of drafting a new oil law taking into account the advice of three oil officials. The new draft, according to Arcaya, was 'adapted to the one I had written . . . reaching in this manner the final draft of the 1922 oil law'.[207] The bill was presented to Congress on 3 June, and six days later on 9 June was approved. Two weeks later Torres was replaced by Antonio Alamo as the Minister of Development.

The new law extended the exploration period from two to three years and contracts were now awarded for 40 years instead of the customary 30 years. Two types of concessions were to be awarded by the government, the exploration–production type, and the production-only type. Under the former, the operator paid an initial exploration surface tax of Bs.0.10 per hectare, and when at the end of the three years it had selected its exploitation blocks it would have to pay another initial production surface tax of Bs.2 per hectare. The production contracts were also liable to this initial tax, and both types of concessions were liable to the following annual surface tax: Bs.2 per hectare over the first three years of the contract, Bs.4 per hectare for the subsequent 27 years, and Bs.5 per hectare for the final 10 years of the contract. Although the new law did not incorporate Torres' financial guarantees, the new tax structure for the production-only concessions compensated for this. Operators with this type of contract paid an effective surface tax of Bs.4 per hectare during the initial year, whereas under Torres' proposals this figure would have been Bs.2.50 (with Bs.0.05 per hectare as a financial guarantee which would be returned later), and then reduced to Bs.0.10 per hectare for the next two years of the exploration period. However, under the new law, surface tax of Bs.2 per hectare per annum would be paid during the first three years, meaning that the government received over the three-year period Bs.8 per hectare instead of the Bs.2.70 under Torres' proposal. The new tax structure was less advantageous to the government when it came to exploration–production contracts, for here the government received during the three-year exploratory period Bs.0.10 per hectare rather than Bs.2.70. Nevertheless, as many concessions had already been awarded to serious concerns, it was likely that these would choose the production-only concession and start exploiting their property as soon as possible. The generous allowance for the exploration–production concession, while allowing speculation to continue, did encourage exploration away from the known oil-bearing lands.

With the new law the companies did achieve a very significant reduction in the overall surface tax rate, something which was of considerable importance to those operators who were expecting to develop the oilfields on a long-term basis. Although the new surface tax rates were much higher than those suggested by Torres, they were nevertheless considerably lower (by 15 per cent) than the rate established by the first law (see Table 2). It was hoped that this loss would be compensated by higher taxes levied on National Reserves.[208] The reduction of the maximum production tax from 15 per cent to 10 per cent of the commercial value of oil can be seen as a further triumph for the oil companies. The 15 per cent royalty, however, had always been artificial since the government, under both previous laws, had pledged to reduce the tax to 10 per cent once two oil-producing wells were situated within 10 kilometres of each other, something which was readily achieved even with small-scale operations. As a result the effective production rate for all serious concerns would invariably be 10 per cent. This provision under the previous laws penalized the companies operating in remote areas. In order to redress this and to stimulate development in the more remote parts of the country, the present law reduced surface taxes by half and production tax by a quarter (to 7.5 per cent) for concessions situated more than 250 kilometres away from the sea or from Lake Maracaibo, or for concessions situated to landward of the Andes.

An examination of the taxes paid by the companies which entered the country during and after 1922, up to 1935, shows that taxes levied per ton produced were very much higher during the initial exploration period than in later years when oil

Table 2 *Comparison between the maximum surface taxes levied by the 1920, 1921 and 1922 Oil Laws, and Torres' proposed Oil Law of April 1922*

	1920	1921	1922 Exploration-production concession	1922 Production-only concession	Torres' proposed Oil Law
Hectares per concession					
Exploration	10,000	10,000	10,000	–	10,000
Production	5,000	5,000	5,000	5,000	5,000
Duration of concession (in years)					
Exploration	2	2	3	–	3
Production	30	30	40	40	30
Initial exploration surface tax (in Bs.)	5,000	5,000	1,000	–	27,000
Initial production surface tax (in Bs.)	25,000	25,000	10,000	10,000	–
Annual production surface tax (in Bs.)	750,000	750,000	820,000	820,000	585,000
Total tax (in Bs.)	780,000	780,000	831,000	830,000	612,000
Tax paid per annum (in Bs.)	24,375	24,375	19,326	20,750	18,545

61

began to flow in great quantities. This pattern is also true when compared to the revenue the country would have received had taxation been levied at 10 per cent of the commercial value of the oil in the U.S. (see Table 3). The revenue achieved per ton of oil produced was considerably higher than has hitherto been acknowledged, and very close to the figure represented by 10 per cent of the commercial value of the oil in the U.S., since writers such as Egaña and Balestrini lumped together taxes paid by companies operating under different taxation schedules. As previously shown, one of the principal reasons for enacting the 1922 Oil Law was to achieve a higher rate of return than that offered by the early contracts held by Shell (which came under the 1910 Mining Code) and which the government recognized as being too generous. In addition, the Bs.20 million paid by Shell in order to retain its concessions[209] have also been included in the tax payments, but this sum should be deducted from the total as it represents special payments to the government and not tax revenue. If production and revenue schedules are adjusted to take this into account, it can be seen (Table 4) that the government during 1922–35 achieved a higher rate of return than has hitherto been recognized. Shell between 1922 and 1935 accounted for 43 per cent of the total production, but only paid 34 per cent of the total tax bill for the period. If it is assumed that Shell paid the same rate of tax for its production, as did the other operators, then the government's revenues for the period would have been much closer to the figure represented by 10 per cent of the commercial value of the oil in the U.S. as advocated by Lecuna and Pérez Dupuy. Here we find that total taxes would have been Bs.509,150,698, of which Shell accounted for Bs.216,743,414, whereas total revenue using the 10 per cent U.S. oil prices scheme would have been Bs.563,655,636, a difference of 9.7 per cent.

Under the new law the companies achieved an outstanding success in gaining exoneration from customs duties. One of the remarkable features of Venezuelan mining law was that it always granted exoneration from customs duties on all the mining equipment brought in by the companies, a practice which was also extended to the oil industry. However, article 48, the relevant clause in the 1922 Oil Law, was so badly drafted that it was open to all sorts of abuse, and allowed the

Table 3 *Revenue received by the government under the 1922, 1925, 1928 and 1935 Oil Laws compared to the theoretical revenue if levied at 10 per cent of the commercial value of Venezuelan oil in the U.S., 1922–1935*

Year	1922, 1925, 1928 and 1935 Oil Laws			10 per cent of commercial value of Venezuelan oil in U.S.	
	Production (in million tons)	Total revenue (in million Bs.)	Revenue per ton (in Bs.)	Total revenue (in million Bs.)	Revenue per ton (in Bs.)
1922	0.004	1.8	405.5	0.02	5.3
1923	0.15	1.7	11.5	0.7	4.4
1924	0.3	2.6	9.0	1.3	4.7
1925	1.5	6.6	4.3	8.4	5.5
1926	2.5	10.6	4.2	15.7	6.2
1927	5.1	11.4	2.2	21.9	4.3
1928	9.7	31.9	3.3	36.8	3.8
1929	12.8	32.9	2.6	53.8	4.2
1930	11.8	28.0	2.4	46.1	3.9
1931	9.8	29.6	3.0	20.7	2.1
1932	11.0	30.0	2.7	31.8	2.9
1933	11.1	29.6	2.7	24.6	2.2
1934	13.5	35.7	2.6	25.6	1.9
1935	15.1	42.4	2.8	28.7	1.9

Notes: figures rounded-off. Exchange rate: Bs. 5.2 = U.S.$1 up to 1933, thereafter Bs.3.09 = U.S.$1. It should be noted that in these calculations average U.S. oil prices have been used. Had U.S. Gulf oil prices, the base rate used by international oil companies in their transactions, been taken, the values expressed for the 10 per cent U.S. oil prices would have been lower because Gulf Oil prices were consistently lower during this period compared to the average U.S. oil prices.
Sources: calculated from The Royal Dutch Company, *Annual Reports* (London, 1907–36); U.S. Tariff Commission, *Petroleum* (War Changes in Industry Series, Report No. 17, Washington, 1946), Table 22, p. 91; Manuel R. Egaña, *Tres décadas de producción petrolera* (Caracas, Tip. Americana, 1947), Table 12, p. 15; César Balestrini, *La industria petrolera en Venezuela y el Cuatricentenario de Caracas* (Caracas, Ediciones del Cuatricentenario de Caracas, 1966), pp. 159–61.

Table 4 *Oil revenues: comparison between Shell Group and other companies, 1922–35*

Year	Production (in tons)			Revenue (in thousand Bs.)					Revenue per ton (in Bs.)	
	Total	% Shell Group	% Others	Total	Shell Group	% Shell Group	Others	% Others	With Shell Group	Without Shell Group
1922	443,700	99	1	2,500	701	28	1,799	72	5.6	405.5
1923	686,841	78	22	2,783	1,078	39	1,705	61	4.1	11.5
1924	1,449,088	80	20	4,914	2,325	47	2,589	53	3.4	9.0
1925	3,163,974	52	48	9,866	3,287	33	6,579	67	3.1	4.3
1926	5,659,438	55	45	16,879	6,245	37	10,634	63	3.0	4.2
1927	9,590,337	47	53	20,428	8,995	44	11,433	56	2.1	2.2
1928	16,818,596,	43	57	46,186	14,251	31	31,935	69	2.7	3.3
1929	21,599,095	41	59	50,544	17,599	35	32,945	65	2.3	2.6
1930	21,467,684	45	55	47,332	19,311	41	28,021	59	2.2	2.4
1931	18,551,334	47	53	46,979	17,425	37	29,554	63	2.5	3.0
1932	18,529,617	41	59	45,145	15,128	34	30,017	66	2.5	2.7
1933	18,761,787	41	59	44,781	15,196	34	29,585	66	2.4	2.7
1934	21,632,909	38	62	52,046	16,337	31	35,709	69	2.4	2.6
1935	23,574,036	36	64	59,297	16,939	29	42,358	71	2.5	2.8

Note: the Bs.20 million paid by Shell to retain its concessions is not included.
Sources: calculated from The Royal Dutch Company, *Annual Reports* (London, 1907–35), U.S. Tariff Commission, *Petroleum* (War Changes in Industry Series Report No. 17, Washington, 1946), Table 22, p. 91; Manuel R. Egaña, *Tres décadas de Producción petrolera* (Caracas, Tip. Americana, 1947), Table 12, p. 15; and César Balestrini, *La industria petrolera en Venezuela y el Cuatricentenario de Caracas* (Caracas, Ediciones del Cuatricentenario de Caracas, 1966), pp. 159–61.

companies to bring in almost everything duty free. According to González Miranda, the only reason for this was to 'establish an exorbitant and manifestly unjust advantage in favour of the producing companies'.[210] The companies imported free of duty such 'equipment' as wood, office materials, paint, desks, doors, medicine, pyjamas, etc.

It is important to compare the benefits accruing to Venezuela as a result of the 1922 Oil Law with the experience of other potential oil producers in order to determine the overall success of Gómez's government. The concessionary system adopted in Venezuela allowed the free entry of all companies, something which the contracts awarded in the Middle East, and the early oil contracts awarded in Venezuela, prevented. A large number of companies operating in the country would result in the government having a greater control of the industry. The Middle East contracts placed the government at the mercy of one company because it had the exclusive right not only to exploit the oil resources of the country, but also to export, refine, and market the oil. Such a monopoly was not allowed in Venezuela; moreover, separate contracts covered the right to refine, transport, and market the oil. The contracts in the Middle East were awarded for the duration of 60–70 years, compared to 40 years in Venezuela. There were, however, similarities, such as the free importation of equipment, and taxation was fixed by the contract and was inviolate. Taxation was also similar, except that royalties were higher, and could be calculated either on the basis of the volume of production, or on profits, or both. Nevertheless, Venezuela during 1913–47 achieved a higher rate of return on its oil than the Middle East countries. During the period in question it received on average $0.23 per barrel compared to $0.21 received in the Middle East. Consequently, the Venezuelan government received a higher percentage of the companies' total receipts than did the Middle East countries (26 per cent compared to 19 per cent).[211] Companies also invested more in Venezuela than in the Middle East (33 per cent compared to 16 per cent)[212] and repatriated less of their net earnings (41 per cent compared to 65 per cent).[213] Production costs per barrel were higher in Venezuela than in the Middle East ($0.52 compared to $0.44),[214] and consequently profits were lower per barrel ($0.66 compared to

$0.88).[215] Indeed, the return per barrel of oil produced was lower in Venezuela than in Mexico (see Table 5).

Table 5   Comparison of revenue (in Bs.) received by Venezuela and Mexico per barrel of oil produced, 1922–35

Country	1922–7	1928–30	1931–5
Venezuela	3.6	2.7	2.8
Mexico	6.9	5.4	3.6

Source: calculated from Lorenzo Meyer, México y Estados Unidos en el conflicto petrolero, 1917–42 (México, El Colegio de Mexico, 1968), p. 3.

The general view of all oilmen, and especially the Americans, was that the 1922 Oil Law was the 'best petroleum law in Latin America'.[216] Exxon, for instance, saw the law as clearing up some very important problems and misunderstandings that had occurred in the interpretation of the previous law.[217] McKay, an extremely active lobbyist, informed Gómez that the law 'protects the interests of the government, while at the same time it is fair to the companies and individuals who invest their capital in the discovery and development of those treasures which nature hides in the soil'.[218] Henry C. Morris, head of the Fuel Department of the U.S. Department of Commerce, suggested in 1922 that other Latin American countries which proposed to exploit their own oil resources should follow the example set by Venezuela.[219] The law was seen, therefore, as a triumph for the oil companies, but the government had scored a remarkable achievement too, especially in view of the vested interests that surrounded Gómez. The new tax structure secured the goodwill of responsible operators by reducing the overall surface and production taxes, and gave the government an immediate injection of oil revenues into the Treasury (one of Torres' principal aims) by allowing companies to acquire exploitation concessions. In addition, the relatively high tax entry fee for these types of concessions would discourage speculators, and thus known oil-bearing lands would not remain fallow.

Nevertheless, the concessions awarded to Gómez's family and friends continued unabated. In June 1922 Eloy Pérez, José Vicente Gómez's attorney, requested certain oil blocks which had already been granted to Juancho Gómez, J.A. Martínez Méndez (Gómez's brother-in-law), Ignacio Andrade (Gómez's son-in-law), and Dr Sosa Altuna. José Vicente acquired the remaining hectareage in the area.[220] In the same year Gómez granted to Pérez Soto, Urdaneta Maya, Colonel Gonzalo Gómez (his son), and Dr Miguel R. Ruiz several oil concessions in Bolívar State.[221]

On 14 December 1922 VOC's well Barroso No. 2 (a well abandoned in 1918 at a depth of 164 metres) erupted for ten days in a manner reminiscent of the famous Mexican gushers, producing 12,000 tons of oil per day.[222] A great deal of this was wasted but eventually 60,000 tons were saved when some 500 labourers dug a hole around the well to contain the oil. The eruption of this well reinforced the view that Venezuela was destined to become a major oil producer.[223]

As it has been shown, Gómez, following his 1908 coup, pursued a strong mining policy that would stimulate industrial development and give his government an independent source of income. The importance of the country's oil resources only became known when the British oil companies started to explore the country in a systematic way after 1913. As a result it was recognized that the Valladares contract had been too generous, and that future oil contracts would have to be more onerous to the companies. But at this juncture there was no certainty that the country would be a major oil producer, and there was a real need to attract investors to the country. For this reason Gómez did not rescind the Valladares contract in 1917 when Hodge brought his claim against it. Moreover, the government showed a great deal of patience with the lack of progress by the British oil companies. In addition to this need, Gómez also wanted the country's oil resources to be developed by companies of different nationalities. Gómez's awareness of the oil development of the country derived from the very close links between his family and the industry. Many also held shares in the companies so that there was a vested interest to see the companies producing oil. The traditional commercial elite shied away from such risky ventures, and so Gómez's entourage, which came close to monopolizing the acquisition of concessions, was able to move into the industry at a very early stage. It would then have been tempting for Gómez to have disregarded the needs of the country and to have pursued a policy that would have benefited only his family and the oil companies. But such a policy had its obvious drawbacks. It would, first of all, create resentment among those loyal supporters who did not receive their share of the oil wealth. Secondly, it would fuel criticism of the government, both internally and externally. More importantly, Gómez's and his government's freedom to act would be constrained by the same traditional economic factors present during his initial term of office, something he wanted to avoid. With healthy oil revenues, the government would be able to undertake projects which would foster the economic progress in the country, and so counterbalance any seeds of discontent. Oil revenues would above all give Gómez a stronger financial basis on which to build his long-term political survival. Consequently, Gómez could not take sides with the *gomecista* concessionaires and with the oil companies which wanted token taxes to be levied, but would have to extract as much tax as the companies could bear. From 1914 onwards the government pursued a consistent policy of securing a larger tax return from the companies. Both the Val de Travers Asphalt Paving Co. Ltd and the Venezuelan Petroleum Exploration Co. Ltd were refused contracts in 1914 because their operating terms were below those offered by the Valladares contract, and in 1916 Díaz Rodríguez made a number of recommendations which were later incorporated in Torres' 1922 Oil Law. These were that higher exploration taxes should be levied, both to give the government an immediate return for the concessions awarded and in order to discourage speculators, and that oil produced should be taxed at a percentage of its value. This policy was retained despite strong pressure from the *gomecista* concessionaires acting on behalf of the companies. The fact that these terms were finally accepted only served to demonstrate what was already apparent during the First World War, which was that oil was of vital importance to the industrialized world, and that new foreign oil sources would have to be found to compensate for the lack of supplies from traditional producers.

It should be noted that the flexibility shown by the government during this period was only possible under a dictatorial regime, where Congress rubber-stamped the Executive's decision. Under a democratic system it is unlikely that the Valladares concession would have been given a second lease of life after the Hodge claim, or that three oil laws would have been enacted in the course of two years. We can, therefore, conclude that Gómez was well informed and aware of the country's oil resources long before 1920; that despite close family connections with the oil industry, he did not become an instrument of the oil companies; that he pursued a consistent policy of exacting higher taxes and that by the end of 1922 the basis of the country's oil industry had been firmly established.

By 1922 the country, and especially Zulia State, was about to embark on an unparalleled economic boom, with all its ensuing political, social, and economic consequences. Having established the oil industry, Gómez would now have to deal with the immense problems created by it.

3

Oil companies and finance

The 1922 Oil Law created the right legal conditions for the oil companies, and from that year onwards Venezuela was to experience an unprecedented increase in oil production, rising from 6,000 barrels per day in 1922 to 425,000 barrels per day in 1936.[1] By 1928 Venezuela had become the second largest oil producer, and the leading oil exporter, in the world, accounting for 8 per cent of total world oil production. Although Venezuelan oil production was dominated by Shell up to 1934, two American companies, Exxon and Gulf Oil, started in 1925 to make inroads into this dominance, and by 1928 had surpassed Shell, controlling 60 per cent of the total production at Gómez's death in December 1935.[2] These three companies accounted for 99 per cent of the total production during Gómez's regime.

Although these three companies dominated the industry, over 120 other companies entered the country to acquire and deal in oil concessions. This large influx of companies, which Torres sought to avoid in order to reduce speculation in the country's natural resources, allowed a large number of people to transfer their oil concessions. As a result, during Gómez's lifetime, the companies acquired 2,434 oil concessions (50 per cent of the total 4,875 concessions awarded to individuals), which had been held by 829 concessionaires (54 per cent of the total number of concessionaires). The companies also acquired a further 1,354 concessions (22 per cent of the total number granted) directly from the government. Consequently, during the period in question, the companies held a total of 3,788 concessions, representing 61 per cent of the total concessions awarded. Not all companies were interested in developing their property, with the result that 22 per cent of the oil contracts held by companies

70

were transferred to other companies. Some made a handsome
profit from this business: for example, in 1926 Alfredo Brandt
acquired 48,000 hectares in Anzoátegui State from Enrique
Silva Pérez for Bs.53,362.[3] Three years later the Venezuelan
American Corp. bought the property from Brandt for Bs.130,625
plus a 2.5 per cent production royalty to him. Soon afterwards,
however, the Caracas Petroleum Corp. acquired it for
Bs.130,000,[4] and later in 1934 sold it to the Orinoco Oilfields
Ltd for Bs.1,149,225.[5] Other speculative companies included
the Maxudian Petroleum Corp., the Venture (Venezuela) Oil
Co. Ltd, the British Venezuelan Syndicate Ltd, the Astor
Petroleum Co., the Venezuelan National Oilfields Corp., and
the National Venezuelan Oil Corp. Others such as the
Pantepec Oil Co. of Venezuela, incorporated in Delaware
(U.S.) in October 1926, and owned by William F. Buckley,
acquired such a large oil acreage that its development was
virtually impossible. In 1927 the company concluded two
separate agreements with the Union Oil Co. of California and
The Texas Corp. for the companies to develop over half the
Pantepec's Oil Co. of Venezuela's property of 3,030,245 acres.
The Union Oil Co. of California, through its subsidiary the
Union National Petroleum Co., would develop 878,000 acres,
investing $3.5 million over the next five years, and profits would
be divided equally between the two companies. The contract
with The Texas Corp.'s subsidiary, the California Petroleum
Corp., was similar except that the company agreed to develop
757,130 acres.[6] Other companies, such as Reitick and Co.,
representing the Consolidated Caribbean Oil Corp., invited
Gómez to invest in the company as a shareholder.[7] There were
other responsible concerns which entered the country with the
intention of developing the oil properties acquired, but which
failed. For example, the Tocuyo Oilfields of Venezuela Ltd, a
British company, acquired from the North Venezuelan
Petroleum Co. Ltd, the oil rights over 85,000 acres at El Mene
in Falcón State.[8] The two companies, which were subsidiaries of
the Central Mining and Investment Corp. Ltd, had invested by
1930 Bs.100 million in developing the property,[9] but the Tocuyo
Oilfields of Venezuela Ltd was not very successful and in 1930
suffered a loss of £131,133, followed by a further loss in 1931 of
£121,566.[10]

Although the number of companies acquiring concessions was large, only a handful came to control more than half of them, and of this handful the two largest American producers, Exxon and Gulf Oil, controlled 43 per cent of all the concessions acquired by companies (see Table 6). As can be seen, many of the companies acquired their concessions from a third party. These intermediaries, individuals who acquired concessions from the original title-holder and then transferred them to either a company or another individual, played the important role of negotiating oil concessions with the companies. Out of a total of 4,875 concessions awarded to individuals during Gómez's regime, the intermediaries held in their own right 6.4 per cent, and negotiated a further 15.2 per cent, managing to transfer over half those held, representing 9 per cent of the total awarded to individuals. The distribution of the concessions held by intermediaries is given in Table 7. The intermediaries have been divided into three groups, the *gomecista* (people close to Gómez who were awarded concessions through his patronage, or who acted as nominees for members of Gómez's family and his entourage), the non-*gomecista*, and the group of foreigners. From Table 7 it is evident that the *gomecista* group, which comprised nearly a third of the intermediaries, accounted for well over half the concessions awarded directly, and for a third of the concessions acquired from original title-holders. However, both the *gomecista* and non-*gomecista* groups were only able to transfer about half the concessions held, whereas foreign intermediaries did much better, transferring 83 per cent of the concessions held. The reason for this was that most of them had close company connections: Warren W. Smith and Henry O. Flipper, for example, worked for the Pantepec Oil Co. of Venezuela. When we look more closely at the real beneficiaries among the *gomecista* group, we find that the Capriles, Sira and Urrutía families, which accounted for 49 per cent of the group, were the most successful. Of the total number of concessions held by the group, the three families acquired 64 per cent, managing to transfer 70 per cent of these, which accounted for a staggering 85 per cent of the total transfers for this group. Table 8 gives the distribution of oil concessions held by the *gomecista* group.

The financial rewards for such activities varied: for example,

Table 6 Distribution of oil concessions held by companies, 1908–36

Holding company	Acquisition of concessions												Number of concessions held					
	D	%	Tr	%	F	%	Tr	%	T	%	Tr	%	Gross	%	Tr	%	Net	%
Exxon	613	45	0	0	471	20	0	0	288	32	0	0	1372	30	0	0	1372	36
Gulf Oil	2	–	0	0	143	6	0	0	100	11	0	0	245	5	0	0	245	6
Sinclair Oil Co.	62	5	216	56	393	17	0	0	42	5	1	1	497	11	0	0	497	13
CVP	304	22	19	5	24	1	17	5	1	–	1	1	329	7	234	28	95	3
Alfred Meyer	37	3			309	13	100	27	102	11	6	7	448	9	125	15	323	9
Pantepec Oil Company of Venezuela	49	4	37	10	133	5	36	10	43	5	21	25	214	5	94	11	120	3
Others (66 companies)	287	21	113	29	904	38	212	58	326	36	56	67	1517	33	381	46	1136	30
TOTAL	1354		385		2366		365		902		84		4622		834		3788	

Key: D concessions acquired directly from the government
F concessions acquired from original title-holder
T concessions acquired from third-party intermediaries
Tr transfers

Note: for the complete breakdown and distribution of concessions see Appendix A
Source: calculated from ADCOTHMEM, 'Historial de Concesiones de Hidrocarburos', Vols. 1–34, Files 1–16,620

Table 7 Distribution of oil concessions held by intermediaries, 1908–36

Intermediaries	Inter-mediaries		Acquisition of concessions												Number of concessions held					
	No.	%	D	%	Tr	%	F	%	Tr	%	T	%	Tr	%	Gross	%	Tr	%	Net	%
Gomecista	38	32	200	63	110	65	251	40	141	38	31	27	7	9	482	46	258	42	224	51
Non-*gomecista*	58	49	112	36	58	35	278	45	151	41	18	15	13	17	408	38	222	36	186	43
Foreign	22	19	4	1	0	0	93	15	78	21	68	58	59	74	165	16	137	22	28	6
TOTAL	118		316		168		622		370		117		79		1055		617		438	

Key: D concessions acquired directly from the government
      F concessions acquired from original title-holder
      T concessions acquired from third-party intermediaries
      Tr transfers

Note: for the complete breakdown and distribution of concessions see appendix A
Source: calculated from ADCOTHMEM, 'Historial de Concesiones de Hidrocarburos', Vols. 1–34, Files 1–16,620

Table 8 Distribution of oil concessions held by gomecista group of intermediaries, 1908–36

Gomecista intermediaries	Intermediaries		Acquisition of concessions												Number of concessions held					
	No.	%	D	%	Tr	%	F	%	Tr	%	T	%	Tr	%	Gross	%	Tr	%	Net	%
Capriles family	5	13	64	32	50	45	45	18	20	14	18	58	1	14	127	26	71	28	56	25
Sira family	2	5	6	3	6	5	48	19	48	34	0	0	0	0	54	11	54	21	0	0
Urrutía family	5	13	23	12	10	9	28	11	27	19	0	0	0	0	51	11	37	14	14	6
Pirella Páez, A.	1	3	52	26	28	26	0	0	0	0	1	3	0	0	53	11	28	11	25	11
Sub-total	13	34	145	73	94	85	121	48	95	67	19	61	1	14	285	59	190	74	95	42
Others	25	66	55	27	16	15	130	52	46	33	12	39	6	86	197	41	68	16	129	58
TOTAL	38		200		110		251		141		31		7		482		258		224	

Key: D concessions acquired directly from the government
F concessions acquired from original title-holder
T concessions acquired from third-party intermediaries
Tr transfers

Note: for a complete breakdown and distribution of concessions see Appendix A
Source: calculated from ADCOTHMEM, 'Historial de Concesiones de Hidrocarburos', Vols. 1–34, Files 1–16,620

in 1930 Adolfo Bueno transferred several oil contracts in Zulia, which he had acquired from CVP for Bs.121,443,[11] to Josefina Revenga de Gómez for Bs.2,800,279,[12] and in 1926 Pedro Augusto Chacín acquired from Sinforoso de Armas a 28,923-hectare concession named Las Bombitas in Anzoátegui State for Bs.21,692,[13] which he then sold in the same year to Henry O. Flipper for Bs.36,154,[14] and who in turn transferred it almost at once to the California Petroleum Corp. for Bs.75,000.[15] The de Armas concession, then, in a matter of months, increased in value by 245 per cent, from Bs.21,692 to Bs.75,000. But other intermediaries were not so fortunate. In 1933, for example, H. Otero Vizcarrondo acquired from Umberto Mondolfi a 6,182-hectare oil property in Anzoátegui State called El Chaparro and Cachipo for Bs.9,000,[16] which the following year he transferred to C.H. Maury for the sum of Bs.9,273, and a 1 per cent production royalty,[17] who in turn sold the concession a few months later to the Caracas Petroleum Corp. for Bs.18,547, and a 1 per cent production royalty.[18]

Although the profit margin for intermediaries varied, the acquisition by them and the oil companies of a large number of concessions had a direct impact on the financial status of the concessionaires. A random selection from the archives of the Ministry of Energy and Mines revealed that local intermediaries paid consistently less for concessions than foreign companies or foreign intermediaries (see Tables 9, 10 and 11). It is interesting to note that in some cases the companies agreed to an annual production royalty, which for Exxon and Gulf Oil amounted to a substantial annual sum because of the large volume of oil produced by these two companies.

During the 1920s negotiations for oil contracts continued at a frenetic pace (abating only slightly with the world economic depression of the early 1930s) allowing almost half the concessionaires to dispose of their concessions and to benefit considerably from this. But this development was also accompanied by the deepening involvement of the Gómez family in oil matters. Moreover, with the incorporation in 1923 of the Compañía Venezolana de Petróleo S.A. (CVP), Gómez, who up to then had stayed out of oil deals, entered the field in a dramatic and pernicious way. The period is also characterized by Gómez's increasing use of oil concessions to secure political

Table 9 *Prices paid by Venezuelan intermediaries for oil concessions, 1925–35*

Year	Title-Holder	No.	Intermediary	Transfer Price Cash (Bs.)	Royalty (%)
1931	Montbrum, J.L.	7	Alamo, C.J.	7,000	
1927	CVP	3	Bueno, A.	228,398	
1931	Howard, O.R.	14	Capriles, R.	260,000	
1926	De Armas, S.	1	Chacín, P.A.	21,692	
1926	Ruiz, A.	1	Chacín, P.A.	9,200	
1930	Bueno, A.	3	De Gómez, J. Revenga	2,800,279	
1926	La Riva Herrera, A.J.	1	Ferris, J.	2,906	
1927	Capecchi, L.	2	López Rodríguez, J.	162,029	1
1927	Ruádez, A. and Poggi, C.	1	López Rodríguez, J.	7,000	
1930	Delgado, F.	2	Madriz, M.	20,000	
1931	Parra, M.	1	De Parra, V. Márquez	10,000	
1925	Briceño Iragorri, O.	1	Márquez Iragorri, J.M.	10,000	
1934	Galavis, H.	3	Martín, F.I.	54,647	1
1934	Itriago Chacín, S.	4	Martín, F.I.	32,335	1
1934	Otero, V., H.	1	Maury, C.H.	9,273	1
1926	José Cervoni Sucs.	1	Mendoza Fleury, L.A.	88,162	
1931	Díaz, M.	1	Otero V., H.	2,390	
1933	Mondolfi, U.	1	Otero V., H.	9,000	
1926	Perea, J.A.	6	Perea, S.	6,000	
1927	Felizola, G.	6	De Tovar, J.M.	178,626	

Key: No.   number of concessions transferred

Note: the Ministry's 'Traspasos' papers only begin in 1925
Source: ADCOTHMEM, 'Traspasos, 1925–35'

Table 10 *Prices paid by foreign intermediaries for oil concessions, 1925–35*

Year	Title-holder	No.	Intermediary	Transfer Price	
				Cash (Bs.)	Royalty (%)
1927	Arcaya, C.	1	Johnston, E.H.	57,402	3
1934	Ferris, J.	1	Muller, E.	3,463	2
1928	Sira, M.A.	30	Porter, E.F.	75,000	
1928	Sira, L. and Corao, M.	36	Porter, E.F.	95,000	
1928	Mendoza, C.L.	1	Robb, E.L.	326,950	
1930	Capriles, I.	?	Schumacher, H.W.	3,030,000	2
1928	Maritime Oil Corp.	3	Shea, J.H.	1,560,000	5
1925	Galán, C. and Pardo, A.	4	Smith, W.W.	150,000	
1926	Tagliaferro, S.	3	Stewart, J.	62,400	

Key: No.   number of concessions transferred
Note: the Ministry's 'Traspasos' papers only begin in 1925
Source: ADCOTHMEM, 'Traspasos, 1925–35'

Table 11 *Prices paid by oil companies for oil concessions, 1925–35*

Year	Title-holder	No.	Oil company	Transfer Price Cash (Bs.)	Shares	Royalty (%)
1932	American Venezuelan Oilfields C.A.	?	Alamo Oil Corp.	171,281		2.5
1932	Venezuelan Pantepec Co. C.A.	?	Alamo Oil Corp.	15,000		2.5
1932	North Venezuelan Petroleum Co. Ltd	?	Alamo Oil Corp.	200,000		
1932	Cia Marítima Paraguaná	?	Alamo Oil Corp.	2,956		
1932	CVP	?	Alamo Oil Corp.	54,735		2.5
1929	Rodríguez Ceballos, H.	2	Alego Oil Concessions	15,035		2.5
1926	Febres Cordero, C.	1	Amerada Corp.	13,854		
1927	García, J.M.	6	American Controlled Oilfields Ltd	2,600,000		5.0
1925	López, G.	1	American Maracaibo Co.	260,000		
1925	García, E.	1	American Maracaibo Co.	208,000		
1925	New England Oil Corp.	1	American Maracaibo Co.	520,000		2.5
1926	Mene Grande Oil Corp.	1	American Marcaibo Co.	38,041		
1926	CVP	1	American Maracaibo Co.	72,800		2.5
1926	CVP	2	American Maracaibo Co.		62,400	
1927	CVP	3	American Maracaibo co.	62,400		2.5
1926	Alfred Meyer	4	Venezuelan Atlantic Refining Co.	234,000		
1927	Griego, J.A.	1	Brokaw, Dixon, and Gardner	26,000		
1929	Mata Contreras, T.	1	Chacin and Lample	9,032		
1929	Otero, V., H.	1	Chacin and Lample	45,680		1.5
1928	Arcaya, C.	1	Cities Service Co.		20,300	
1929	Arcaya, C.	1	Cities Service Co.	96,343		
1935	Venezuelan Pantepec Co. C.A.	1	C.A. de Petróleo Altagracia	10,000		3.0
1935	Pantepec Oil Co. of Venezuela	?	C.A. de Petróleo Altagracia	890		2.5
1935	Alamo Oil Corp.	?	C.A. de Petróleo Altagracia	1,755		1.0
1925	Reyes Rivero, F.	1	CVP	100,000		1.0
1926	Fernández, A.	1	CVP	49,966		

Table 11 (*cont.*)

Year	Title-holder	No.	Oil company	Transfer Price Cash (Bs.)	Shares	Royalty (%)
1926	Tovar, J.M.	1	CVP	34,350		
1926	Urrutía, C. et al	18	Sinclair Oil Co.	728,000		
1927	Tirado Medina, A.	1	Sinclair Oil Co.	2,000		
1927	Carvallo, L. and E.	10	Sinclair Oil Co.	210,000		
1935	Chiara, U.	4	Sinclair Oil Co.	80,993		2.5
1929	D'Empaire, P.A.	6	Dakota Oil and Transport Co.	33,333		
1929	Loinaz, A.	6	Dakota Oil and Transport Co.	33,333		
1927	CVP	?	Falcon Oil Corp.	210,080		2.5
1927	CVP	?	Falcon Oil Corp.	215,280		2.5
1928	CVP	?	Falcon Oil Corp.	197,600		2.5
1935	Chiara, U.	27	General Asphalt Co.	252,218		2.5
1925	Urdaneta Exploration Co.	?	Gulf Oil	102,798		
1925	Alfred Meyer	6	Gulf Oil	390,000		
1926	Miranda Exploration Co.	?	Gulf Oil	143,910		
1926	Perijá Exploration Co.	?	Gulf Oil	104,000		
1926	Sucre Exploration Co.	?	Gulf Oil	199,212		
1927	Kunhardt and Co.	3	Gulf Oil	103,800		
1927	CVP	6	Gulf Oil	567,011		2.5
1930	Brice, A.F.	?	Gulf Oil	5,000		
1931	Berg, N.	?	Gulf Oil	26		
1933	Ferris, J.	8	Gulf Oil	5,000		
1933	Egaña, M.R.	1	Gulf Oil Corp.	30,000		
1927	Pietri, J.	1	Edwin B. Hopkins	240,000		2.5
1927	Ruiz, J.	4	Edwin B. Hopkins	27,097		2.5
1926	Urrutía, L.A. and Capriles, J.E.	18	Intercontinent Petroleum Corp.	197,600		
1930	Caballero, L.	27	Lagomar Oil Concessions		999,000	
1925	Werner, C.	2	Alfred Meyer	193	14,875	

1926	CVP	Alfred Meyer	1	12,969		1.0
1929	Algeo Oil Concessions	Alfred Meyer	17		314,000	1.0
1934	Maury, C.H.	Alfred Meyer	1	18,548		2.0
1934	Muller, E.	Alfred Meyer	1	3,000		5.5
1928	Shea, J.H.	George F. Naphen	3	4,992,000		2.5
1929	Bance, J.B. and Chacín, P.	National Venezuelan Oil Corp.	1	15,000		2.5
1933	Algeo Oil Concessions	Orinoco Oilfields Ltd	1	910,000		
1925	Jiménez Arraiz, F. and Ramírez R., A.J.	Pantepec Consolidated of Venezuela Inc.	12	624,000		
1925	Stelling, O.	Pantepec Consolidated of Venezuela Inc.	2	26,000		
1925	González, J. and Pinto, S.	Pantepec Consolidated of Venezuela Inc.	2	126,000		
1926	Smith, W.W.	Pantepec Consolidated of Venezuela Inc.	5	75,000		
1926	CVP	Pantepec Consolidated of Venezuela Inc.	14	25,000		
1926	CVP	Pantepec Consolidated of Venezuela Inc.	?	2,593,000		
1926	New England Oil Corp.	Pantepec Consolidated of Venezuela Inc.	2	134,254		
1927	López Rodríguez, J.	Pantepec Consolidated of Venezuela Inc.	9	75,075		
1928	CVP	Pantepec Consolidated of Venezuela Inc.	11	7,418		
1928	González, E.	Pantepec Consolidated of Venezuela Inc.	1	12,500		
1931	Union National Petroleum Co.	Pantepec Consolidated of Venezuela Inc.	6	29,000		
1927	Molinos, J.	Seaboard Oil Co. of Delaware	5	1,348,631		3.0
1927	CVP	Seaboard Oil Co. of Delaware	?	38,381		2.5
1926	CVP	Simms Petroleum Co.	?	211,868		
1929	Urdaneta Exploration Co.	SOC(California)	1	64,246		
1925	Abenum de Lima, M.	Exxon	5	22,753		
1925	Alfred Meyer	Exxon	5	312,000		
1925	Iturbe, G.	Exxon	2	265,000		
1925	Urrutía, C.	Exxon	3	260,000		
1925	Etanislao Arcaya, E.	Exxon	1	42,712		
1926	Alfred Meyer	Exxon	31	1,560,000		2.5
1926	Capriles, M.J.	Exxon	3			
1927	Jordan, W.C.	Exxon	5	75,000		
1927	Wanzer, C.H.	Exxon	5	71,000		
1927	Ware, C.B.	Exxon	3	47,000		
1928	Alfred Meyer	Exxon	35	1,100,000		

Table 11 (*cont.*)

| Year | Title-holder | No. | Oil company | Transfer Price | | |
				Cash (Bs.)	Shares	Royalty (%)
1930	Chiara, U.	9	Exxon	1,554		
1930	De Capriles, A.	4	Exxon	29,408		2.5
1933	Pantepec Oil Co. of Venezuela	15	Exxon	5,000		
1934	Antonio Díaz Oilfields Ltd	9	Exxon	180,000		2.5
1934	De Tinoco, J. Revenga	?	Exxon	2,000		
1935	Pantepec Oil Co. of Venezuela	8	Exxon	5,200		2.5
1935	Alamo Oil Corp.	12	Exxon	27,368		2.5
1935	Chiara, U.	36	Socony-Vacuum	91,846		3.0
1935	Martín, F.I.	10	Socony-Vacuum	32,335		
1927	Pantepec Oil Co. of Venezuela	22	The Texas Corp.	438,000		
1927	Flipper, H.O.	1	The Texas Corp.	75,000		2.5
1928	Smith, W.W.	1	The Texas Corp.	273,000		
1930	New England Oil Corp.	3	The Texas Corp.	5,200		
1927	Venezuelan Oil Corp.	2	Union Oil Co. of Venezuela	83,000		
1927	Cia Marítima Paraguaná	5	Union Oil Co. of Venezuela	44,000		
1927	Pantepec Oil Co. of Venezuela	15	Union Oil Co. of Venezuela	179,000		
1927	Pantepec Oil Co. of Venezuela	?	Union Oil Co. of Venezuela	292,000		
1927	Piñerúa, G.	2	Venezuela Syndicate	110,000		2.5

Key: No.   number of concessions transferred
Note: the Ministry's 'Traspasos' papers only begin in 1925
Source: ADCOTHMEM, 'Traspasos, 1925–35'

loyalty and to reward faithful government officials. A more detailed study of the contracts made with private individuals who sought Gómez's help, and of those made with government officials, and a study also of the role played by Gómez and his family and close entourage will give a clearer picture of the deals struck at the time.

As the great financial benefits accruing to the oil concessionaires became apparent, requests to Gómez for oil concessions increased: in 1923 David Rodríguez requested an oil concession,[19] the following year F. Conde García sought two concessions of 15,000 hectares each,[20] and in 1926 M.S. Briceño requested a concession of 30,000 hectares.[21] Most were unsuccessful in their requests, but others such as Jose León were successful.[22] Others, such as General Arturo Uzlar and his wife Doña Elena, used their influence with Gómez and Baptista Galindo, his Secretary-General, to acquire concessions, requesting in 1926, for example, that he be granted two National Reserve blocks created by the Bolívar Exploration Co. Ltd in the Delta Amacuro Territory.[23]

Those who could not afford to obtain oil titles on their land would offer Gómez a share in the future profits of the concession if he would agree to waive the initial taxes and legal fees associated with the concession. Rafael A. Hermoso, for instance, offered Gómez such a consideration in 1924 when he proposed to grant him 70 per cent of the net profits realized over and above the value of his land (Bs.100,000) from the transfer fee of the oil titles he was to acquire in Zulia, and as he expected a transfer fee of between Bs.1,040,000 and Bs.1,560,000, Gómez's share would amount to between Bs.728,000 and Bs.1,092,000.[24]

Other people who acquired concessions, but who were later involved in financial or legal problems, appealed to Gómez for help. In December 1927, for example, Mrs Dolores L. de Bello Rodríguez acquired an 84-hectare oil concession entitled El Ciclón which was made up of National Reserves in the Mara District of Zulia State. The transaction was negotiated through CVP to compensate for it having involuntarily acquired a concession already granted to Mrs Bello.[25] She thought that CVP would pay all legal costs involved in the transaction, as it had done in the case of two other concessions. However, enquiries at the Ministry in 1929 revealed that the transfer deeds had not been presented, and that there was therefore a strong possibility that the concession would be rescinded if the Minister did not agree to the presentation made now.[26] However, in February 1930 it was agreed that the concession would be allowed to lapse but that the Minister would grant a similar one later on. When the new title deeds were not awarded, Mrs Bello appealed to Gómez through Colonel Sixto Tovar, then his Secretary-General, to find a solution. Another example of this type of appeal was made by Laureano Vallenilla Lanz on behalf of his mother-in-law, Doña Carmen Lecuna de Blanco. In 1929 she had acquired an oil concession in Barlovento, which was registered under the name of General Mata Meneses, but which by 1934 was about to expire. Doña Carmen appealed to Gómez for help.[27] He acceded to this request, promising to order Cayama Martínez, the Minister of Development, to waive 'the fees to be paid for the "Blanquillo" and "Potrero de Ponte" concessions'.[28] However, the situation remained unchanged up to 1934, and Vallenilla Lanz appealed to Urdaneta Carrillo, Gómez's Secretary-General at that time to find out the cause of the delay. Ubaldo Chiara was another who wanted the government to write off the back taxes he owed. Between 1929 and 1930 Chiara had acquired a large number of concessions, spending Bs.300,000[29] in legal fees, topographical surveys, etc., but had been unsuccessful in disposing of them. In 1935 he had received firm offers from the Atlantic Refining Co., Sinclair Oil Co. and Socony-Vacuum to purchase several of them on the condition that the taxes owed had been paid in full. As a result Chiara appealed to Gómez for the government to write off the taxes (although he was willing to pay up to 10 per cent of the taxes owed once the transfers had taken place) because it would be in the government's interest to change a 'poor debtor for a rich one'.[30]

Jacinto López was another who benefited from Gómez's help. With Gómez's influence he had acquired a small concession which he was able to transfer to the Venezuelan Pantepec Co. C.A., after which he retired to Paris.[31] In late 1931 he received the bad news that the company was about to renounce his concession because it had been unsuccessful in finding oil, preferring to give it up rather than to pay taxes on it. As a result López decided to renounce the other concessions which he held because he did not want to increase his debt with the government.[32] To achieve this, he wanted Gómez to exonerate him from all the taxes which he owed. Later in 1932, however, López acquired a further concession called Santa Ana de el Lirial, next to four concessions which Exxon had obtained from Venezuelan Pantepec Co. C.A. in Monagas State.[33] Three years later he suggested to Urdaneta Carrillo, then Gómez's Secretary-General, that he procure the National Reserves on this block because he could negotiate with Exxon a production contract by which Gómez would receive half the total oil produced.[34] That same year López appealed to Gómez to grant him the titles to a concession called Apurito which he had originally held, but which he had later renounced with the hope of reacquiring it. He now had a firm offer from Socony-Vacuum for the concession, and in addition would secure a 2.5 per cent production royalty.[35]

Others such as Lucio Baldó (the former Technical Director of CVP) who had incurred large debts appealed to Gómez for oil concessions to cover their debts. In this case, Baldó had offered to sell his property of Las Adjuntas to the nation because the banks which had lent him money were recalling their loans.[36] He therefore wanted Gómez to grant him a few oil concessions which he could then negotiate and use the proceeds to cover his debts.

GOVERNMENT OFFICIALS

As has been shown, government officials were barred by law from obtaining oil concessions. However, it was not illegal for them to obtain concessions after leaving office, or as landowners during the period when the 1920 Oil Law was in force. Table 12 shows both the former and the serving government officials who obtained concessions in their own names. The majority of the concessions given in Table 12 were acquired legally, and it is surprising that relatively few government officials obtained oil concessions in their own right. Perhaps the answer is a combination of poor oil-bearing lands, and overambitious transfer prices. For the latter there is evidence that the officials tried hard to transfer their concessions. For example, Santiago Fontiveros, the Minister of Development 1914–16, and

Table 12 *Former and serving government officials who obtained oil concessions in their own names, 1908–35*

Name	Government position held	Acquisition of concessions		
		Year	No.	Hectareage
Alvárez, R.R.	Public Works Minister, 1910	1921	6	60,000
Alvárez Feo, F.	Public Works Minister, 1929	1921	6	60,000
Andrade, I.	Interior Minister, 1917	1925–9	10	100,000
Arcay, J.F.	President, Cojedes State, 1915–24; President, Carabobo State, 1924–7	1922	3	11,155
Barreto Méndez, G.	President, Cojedes State, 1924–7	1927–8	2	12,361
Blanco, L.F.	Senator	1921	2	20,000
Carvajal Madrid, B.	Senator	1925	1	8,800
Chalbaud Cardona, E.	President, Mérida State, 1909–13	1921	6	60,000
Dávila, C.A.	Senator	1927	6	60,000
Díaz González, M.	Public Works Minister, 1935	1922	6	60,000
Díaz Rodríguez, M.	Foreign Minister, 1914–16; President, Nueva Esparta State, 1925; Senator, 1918	1922	1	5,685
		1928	1	1,494
Febres Cordero, I.	President, Zamora State, 1909–14, 1921–4; President, Zulia State, 1925	1925	4	34,876
Fontiveros, S.	Development Minister, 1914–16; President, Trujillo State, 1915–18, 1921–4	1923	1	571
Gabaldón, J.J.	President, Apure State, 1909–14; President, Monagas State, 1929; President, Anzóategui State, 1930	1922	2	7,734
Galavís, F.	President, Aragua State, 1909–13	1925	4	136,336
	President, Yaracuy State, 1929	1929	1	146,953

Name	Position	Year		Amount
García, J.M.	President, Zulia State, 1915–18	1911	1	300
	President, Carabobo State, 1928	1921	6	60,000
Gil Fortoul, J.	Governor, Federal District, 1929	1928	1	10,000
	Education Minister, 1911	1927	10	100,000
	Provisional President, 1913–14			
Gimón, D.	President, Lara State, 1914–21	1926	2	139,349
Gómez, J.C.	President, Lara State, 1909–13	1922	4	14,051
	Governor, Federal District, 1913–23			
González, B.	Senator	1927	1	300
Guerra, P.M.	Governor, Federal District, 1913	1923	1	7,500
		1930	2	16,500
Gutiérrez, F.	Senator	1922	2	950
Hidalgo, J.	President, Aragua State, 1914–18	1923	1	–
	Governor, Federal District, 1923			
Jiménez, J.V.	President, Yaracuy State, 1909–13	1922	1	3,983
Maldonado, S.D.	Education Minister, 1908–10	1921	1	462
	President, Aragua State, 1921–4	1922	3	63,603
	Deputy			
Niño, S.E.	Education Minister, 1929	1928	4	40,000
		1933	12	120,000
Pérez Limardo, J.A.	Senator	1925	1	214
Pérez Soto, V.	President, Portuguesa State, 1909–13	1929	1	97,709
	President, Bolívar State, 1914			
	President, Apure State, 1921–4			
	President, Bolívar State, 1925			
	President, Zulia State, 1926–35			
Pietri, J.	Foreign Minister, 1909	1922	6	60,000
		1927	1	2,310
Pimentel, A.	Finance Minister, 1910–12	1928	1	50,000
Planas, B.	Finance Minister, 1910–11	1907	1	–

87

Table 12 (*cont.*)

Name	Government position held	Acquisition of concessions		
		Year	No.	Hectareage
Rodríguez Gonzalez, E.	Secretary General to Provisional President, 1918	1920	3	45,000
Rolando, A.	Senator	1922	1	6,716
Ramírez, J.A.	President, Nueva Esparta State, 1909–21 President, Sucre State, 1924–7 President, Táchira State, 1925 President, Guárico State, 1929	1922	1	251
Tagliaferro, J.A.	Senator	1922	1	1,000
Terius, D.A.	Senator	1922	1	10,000
Torres García, M.	President, Bolívar State, 1915–21	1922	3	12,512

Source: adapted from ADCOTHMEM, 'Historial de Concesiones de Hidrocarburos', Vols. 1–34; *DDCS, DDCD,* 1913–15; and Pedro Emilio Fernández, *Gómez, el Rehabilitador* (Caracas, Jaime Villegas Editor, 1956), pp. 327–39.

President of Trujillo State 1916–18 and 1921–4, obtained on 3 August 1923 an oil concession covering some of his property in Trujillo State which was adjacent to a concession held by the Venezuelan Sun Ltd and the Standard Oil Company of Venezuela (SOC (Venezuela)). Fontiveros was assured by José Antonio Tagliaferro, José Tomás Carrillo Márquez, and a geologist named Evanoff that his concession was situated on good oil-bearing land. As a result Pedro Rafael Rincones jun. in New York offered Fontiveros' property to Louis B. Wekle for the sum of $300,000.[37] Wekle, however, considered the price steep and later procured several National Reserves from CVP.[38] All the other officials who acquired concessions followed the same course as Fontiveros. Bernabé Planas and David Gimón were the only exceptions, the former transferring in 1915 his concession to the Venezuelan Fuel Oil Syndicate Ltd, and the latter receiving a purchase option of Bs.242,000 from the Pantepec Oil Co. of Venezuela.[39]

Former officials who were in serious financial difficulties also appealed to Gómez for help. In 1935, for instance, G.T. Villegas Pulido, the former Attorney-General, owed the Banco de Venezuela Bs.100,000, and requested from Gómez two concessions of 20,000 hectares each in order to sell them to pay off his creditors.[40] Some family members of former government officials also benefited from Gómez's generosity: for example, when Dr Abel Santos, the former Finance Minister, died, Gómez granted his brother Eduardo E. Santos an oil concession situated in the Delta Amacuro Territory, which he managed to transfer to SOC (Venezuela) in 1933.[41]

The interest which government functionaries had in oil matters did not end there. There were several important transactions and negotiations which public servants in office entered into with the oil companies with the full knowledge and approval of Gómez. In 1922 Gómez granted to V. Pérez Soto, together with Dr Enrique Urdaneta Maya, Colonel Gonzalo Gómez (Gómez's son), and Dr Miguel R. Ruiz 'several oil blocks in the lower Orinoco'.[42] Pérez Soto was the principal partner in the syndicate formed which for legal reasons was headed by Néstor Maya. Pérez Soto incurred all the expenses in his efforts to transfer the concession, and in 1926 A.J. Ramírez Román took out an option on the concession, which he exercised in January 1927, acquiring half of it for which the partners received Bs.27,500 each. In 1925 the Cities Fuel Power Co., a company controlled by Henry L. Doherty,[43] acquired from Pedro M. Arcaya, the Minister of the Interior, an option on 780,000 hectares in Zamora State. The terms of the option were the following: Arcaya would receive a cash payment of Bs.10,000 immediately, followed by $15,000 on 15 May 1926; the company had to acquire a minimum of 250,000 hectares at $0.75 per hectare, i.e. $195,000. Another 520,000 hectares could be obtained for $156,000, and Arcaya retained a 2 per cent production royalty.[44] Antonio Alamo, the Minister of Development, was also involved in illegal pacts with the British Controlled Oilfields Ltd. The company was informed by Alamo's predecessor on 8 May 1922 that the government intended cancelling its concession over the Buchivacoa District of Falcón State (except for the eight blocks the company had in production) because of its failure to develop the property adequately. After lengthy negotiations, which were complicated by Exxon agreeing to develop part of the company's property, the government on 31 May 1925[45] agreed to allow the company to retain its concession, and recognized that the company's concession came under the 1910 Mining Law, thus halving its taxes. Moreover, it allowed it to transfer part of its concession to another company. Although the agreement had the external appearance of being a case hard-won by the company, Alamo appears to have been bribed by the company to accept its terms, for under a separate, secret, agreement the company transferred 50,000 shares to Alamo's New York bank account with the National City Bank of New York in October 1924. Two weeks later, on 4 and 6 November, he transferred 20,000 shares to an anonymous purchaser for the sum of Bs.486,850.[46] Alamo further benefited from his official position when in August 1925 Gómez distributed Bs.1 million among his Cabinet,[47] and in 1928 granted 1 million hectares of oil concessions in Zamora State to seven members of his Cabinet who included Alamo.[48] Isaac Capriles was nominated the intermediary to negotiate the concession. However, because the concessions were granted in one block and not to individual Ministers, Alamo later confessed that 'we quarrelled over the negotiations, each going his own way, and all emerging at the end worse off'.[49] But in 1933 Gómez agreed that the Ministers could reacquire their share individually. As a result Pedro M. Arcaya, José Ignacio Cárdenas, and Jiménez Rebolledo (whose titles appeared under Manasés J. Capriles' name) took up the offer. Later, in March 1935, Alamo wanted to accept the offer because he had a proposition from a foreign company to acquire 140,000 hectares for Bs.280,000 and a 2.25 per cent production royalty. The intermediary who negotiated the contract demanded Bs.84,000 for his services, and Alamo was prepared to offer Urdaneta Maya, Gómez's Secretary-General, an equal amount if the concession was awarded to him.[50]

It should be noted, however, that few government officials were involved in oil deals. The Jurado de Responsabilidad Civil y Administrativa set up in 1946 to investigate the illicit earnings of government officials during the Gómez dictatorship, and during the subsequent two administrations of General E. López Contreras and General I. Medina Angarita, uncovered remarkably little evidence to support the charge of fraudulent earnings from oil. The illicit earnings made by government officials came chiefly from running illegal state monopolies, from real estate deals financed partly from payments received from the infamous Capítulo Séptimo of the Interior Ministry, and from running various State *remates*.

OIL DEALS AND GOMEZ'S FAMILY

Nevertheless, it was Gómez's close family and entourage that would benefit most from the development of the oil industry during the 1920s. As we have seen, the Capriles family formed one of the largest and most successful groups of *gomecista* intermediaries. On 4 February 1926 Preston McGoodwin, the former U.S. Minister at Caracas, and the representative of Isaac Capriles, sold to Sinclair Oil Co. 2,460,000 hectares for Bs.947,887 cash, 5 per cent of the shares in the new company to be formed, and a 1.5 per cent production royalty.[51] On 3 March 1928 Isaac Capriles obtained a further 1 million hectares in Apure State which he transferred to two subsidiaries of Sinclair Oil Co., the Apure-Venezuelan Petroleum Corp. and the Zamora Venezuelan Petroleum Corp.[52] Dr Adolfo Nouel, Gómez's dentist and confidant, acquired three concessions which in 1928 he transferred to a Franco-Hispanic syndicate, Conde Moltke Cianosoff & Associates, in return for a 45 per cent production royalty.[53] Dr Adolfo Bueno became Gómez's personal physician in the early 1920s, and later acted as his private secretary. On 14 June 1922 he requested Gómez to grant Gustavo Escobar Llamozas three oil concessions, which were later transferred to the British Equatorial Oil Co. for a 5 per cent production royalty to be divided as follows: Adolfo Bueno, 65 per cent; Gustavo Escobar Llamozas, 20 per cent; Tomás Duarte P., 10 per cent; and Carlos Hey, 5 per cent.[54] Bueno received Bs.2 million in royalty payments before the Falcón Petroleum Corp. (an American company) on 20 July 1927 procured the 5 per cent production royalty for Bs.14,040,000[55] of which Bueno's share was Bs.9,126,000.[56]

Others with closer family ties also benefited. Among these were General F.A. Colmenares Pacheco, married to Gómez's sister Emilia; General J.A. Martínez Méndez, married to Gómez's sister Indalacia; and Santos Matute Gómez, Gómez's cousin. In August 1923 Colmenares Pacheco sold a kilometre-wide strip of 6,858 hectares over Lake Maracaibo to the Venezuelan Gulf Oil Co. for Bs.473,200[57] and a 5 per cent production royalty.[58] In 1922 Juan E. Paris jun., a prosperous merchant from Maracaibo, proposed to General Martínez Méndez that he apply to Gómez for National Reserves covering 150,000 hectares in Zulia State. The concessions would be extremely valuable because they were composed of land which had been formerly held by companies in Zulia. Under the agreement the concessions would be granted to R. Isava Núñez, Secretary-General to Martínez Méndez, but the General would acquire 75 per cent of the property, and Paris the remaining 25 per cent.[59] Gómez approved the plan and in 1922 Isava Núñez obtained the National Reserves created by the Maracaibo Oil Exploration Co. group of companies in Zulia.[60] Isava Núñez immediately granted Juan E. Paris jun. and the Sociedad Paris & Cia of New York power of attorney to negotiate and transfer the concessions to an American company. On 22 October 1922 a contract was entered into between Paris and the Pure Oil Co. to develop and exploit the concession. A new company, the Orinoco Oil Co., was formed and registered in Delaware with an authorized capital of $4 million in $100 shares, of which $1 million was fully paid up. The Venezuelan concessionaires would receive one share per 15 hectares transferred, i.e. a total of 10,000 shares in the new company, representing $1 million.[61] In 1925 Santos Matute Gómez solicited from Gómez the subsoil oil rights to his Zulia farms of Tasajeras, Ciénaga de Lagunilla, and Purgatorio.[62] In 1928 he sold his Tasajeras del Norte property to VOC for Bs.4 million[63] and the remaining quarter, which belonged to his daughter Ana Beatriz, was also sold to the company.[64] In 1930 Angel Francisco Brice, acting on behalf of Santos Matute Gómez, applied to Torres for 57 concessions in Anzoátegui State, which Torres refused to award before presidential approval had been obtained. Consequently, Brice and Octavio Calcaño Vethancourt, another lawyer employed by Santos Matute, went to Maracay and explained their position to Gómez's Secretary-General, who after consulting Gómez 'confirmed that the matter was settled with the General'.[65] As a result, Brice received between May 1930 and April 1931 76 oil concessions in Anzoátegui State,[66] of which nearly three-quarters were transferred to the Orinoco Oilfields Ltd, a British company.

The *gomecista* concessionaires who had negotiated some of the largest oil concessions prior to 1920 were also prominent at this stage. Julio F. Méndez, for instance, on 23 June 1922 sold to Kunhardt and Co. a kilometre-wide strip over Lake Maracaibo which ran from Lagunillas to La Ceiba.[67] In 1925 Gulf Oil acquired 1,300,000 hectares in Anzoátegui State from Addison H. McKay,[68] and Ignacio Andrade sold to Harold J. Wasson a number of National Reserves in Sucre State for Bs.260,000,[69] and in 1924 40,000 hectares of National Reserves, also in Sucre State, to Exxon for Bs.800,000.[70] At the same time Rafael Requena together with Manuel Alvarado, his brother-in-law who was a naturalized Frenchman, were instrumental in attracting French oil interests to the country.[71]

Up to the time of his murder in June 1923, Juancho Gómez, too, increased his involvement in oil matters. In 1920 he informed Torres through A.J. Ramírez Román and F. Jiménez Arraiz that he would grant Torres 10 per cent of the proceeds of several blocks in which he was interested,[72] and later Ramírez Román agreed to hand over to Juancho Gómez one-sixth of the 33 per cent share of the proceeds which Oswaldo Stelling paid to Ramírez Román from the sale of his Ojo de Agua and Corepano concessions.[73] Juancho was also involved in secret partnerships with other *gomecistas*. He and F.A. Colmenares Pacheco, Márquez Bustillos, and José Antonio Tagliaferro owned 400,000 hectares of oil lands in the Perijá District of Zulia, which were transferred to the Venezuelan Atlantic Refining Co.[74] Juancho also entered into an agreement with General Ulpiano Olivares to acquire oil concessions. Juancho did not live to see this deal come to fruition, but in 1924 Olivares acquired 28,000 hectares which he transferred to the New England Oil Co. for a cash payment of $12 per hectare and a 2.5 per cent production royalty.[75] Lastly, as previously shown, Juancho had acquired from Valbuena, Espina and Bohórquez a 25 per cent stake in the outcome of their legal case against CPC for possession and damages to four asphalt mines they had acquired in 1904, and which the Valladares concession covered. After a long, drawn-out case which lasted over nine years,[76] an agreement was reached on 4 February 1924 whereby the parties concerned would sell the property under dispute and four other oil blocks belonging to the company to a third party and divide the proceeds equally among them. A Comisión Vendedora, composed of Rafael Max. Valladares and Gustavo Nevett, was set up to dispose of the property, which in September 1927 was sold to the Rio Palmar Oilfields Exploration Corp.[77] (a subsidiary of Exxon) for Bs.3,380,000,[78] with Juancho's share of the proceeds being Bs.422,500.

General José Vicente Gómez, the Vice-President and the Inspector-General of the Army up to 1928, acquired in 1921 a vast holding of 1,159,000 hectares over Lake Maracaibo. The concession was to act as the focal point and main attraction to companies which would co-operate in developing a large industrial, agricultural, and commercial scheme in Zulia which José Vicente planned together with Carlos Delfino. Consequently, as soon as the titles were published in the Official Gazette on 16 August 1922, Delfino left for the U.S.[79] to negotiate with a number of oil companies. In addition, Delfino was to negotiate a contract with a company for the construction of an oil pipeline between Colombia and Venezuela. In promoting his project Delfino was assisted by J.A. Coronil, Venezuelan Vice-Consul at New York, Preston McGoodwin,[80] and the ubiquitous Addison H. McKay. These last two would receive a share of the proceeds from the negotiation as payment for their services. Delfino's plan was to organize an American–Venezuelan Development Corp. comprising several different companies which were to co-operate in developing 'industries in general, agriculture, commerce, breeding and cattle raising, mining and the petroleum industry in its various phases and ramifications etc.'.[81] Dr Van H. Manning of the American Petroleum Institute was first consulted, and later the project was discussed with Wallace Mein of the South American Gulf Oil Co. who declared 'himself to be frankly of the opinion that co-operation on all development in Venezuela could be obtained with the exception of oil'.[82] Dr Manning also took up the matter with Walter C. Teagle, the President of Exxon, who thought initially that 'it was possible to form a development company with the exception of oil'.[83] Nevertheless, after further discussion, Sadler, Vice-President of Exxon, informed Delfino that his company was willing to enter into such a development corporation provided that it acquired all the National Reserves arising from the present and future concessions. Such an agreement was unacceptable to Delfino, and it appeared that the oil fraternity was unwilling to co-operate in a joint venture. Moreover, the concession itself was unattractive, for new techniques would need to be developed in order to drill over the shallow waters of the lake. But Edward L. Doheny, owner of the Pan American Petroleum and Transport Co. which through its subsidiary, the Colombian Petroleum Co., operated the Barco concession in Colombia, became interested in Delfino's project, and also in the proposal for a pipeline between Colombia and Venezuela, for such a pipeline would considerably reduce the cost of exporting Colombian oil. Moreover, Doheny was thinking of transferring his large Mexican oil operations to Venezuela. As a result J.A. Coronil held several meetings with Doheny's staff at the end of which, in April 1923, it was decided that the company would take out an option for the oil concessions covering Lake Maracaibo.[84] The concessions were to be held by a new company formed by Preston McGoodwin, J.S. Cosden, and Payne Whitney Associates called the Lago Petroleum Co. (Lago), which was registered in Delaware on 12 April 1923 with an authorized capital of 4 million shares of no par value, of which 2,992,500 were fully paid up.[85] Doheny also entered into an agreement with Claudio Urrutía, the brother-in-law of Pedro Manuel Arcaya who was the Venezuelan Minister at Washington, to acquire several oil concessions which he was negotiating.[86] So in May 1923 Urrutía with Walker and Archer (both Vice-Presidents of the company), J.J. Cotter, and a group of geologists travelled to Venezuela on board the yacht *Casiama* to inspect the property.[87] After they had submitted their report Doheny decided to purchase the concession for Bs.18,750,000[88] and 1,399,970 shares in Lago[89] which later became part of Exxon. Delfino was less successful in negotiating the other concessions which he held jointly with José Vicente. He had formed a company which owned these concessions, and although he was aware that these were not valuable he believed that if they were combined with some other more valuable concessions a good bargain could be struck.[90]

Although José Vicente's development project did not materialize he persisted with it, albeit on a less grand scale, when his wife, Josefina Revenga de Gómez, became the sole owner of the C.A. American Investment Co. According to the Jurado de Responsabilidad Civil y Administrativa, this company held 300 shares in the Banco Mercantil y Agrícola, 50 shares in the Sindicato de Caracas, 11 debentures of Bs.5,000 each and 25 debentures of Bs.1,000 each in the C.A. Luz Eléctrica de Venezuela, and several buildings and houses.[91] José Vicente also established another company, the C.A. Venezuelan Development Corp., with an authorized capital of 1,600 shares of Bs.1,000 each, and 89 debentures of Bs.10,000 each, which was solely owned by his children.[92] The company held property interests as well as 120 shares in the Banco Mercantil y Agrícola, and 8,204 shares in C.A. Luz Eléctrica de Caracas. José Vicente continued negotiating concessions up to his death in Switzerland in 1930 of acute diabetes.[93] José Vicente's sister, Flor María Gómez de Cárdenas, also held several concessions under various nominees.[94]

The illegal business deals entered into by Gómez's family and intimates were small compared to Gómez's own business deals in oil. Offers of concessions for sale and shares in companies were constantly received by him; for instance, in 1928 Pedro F.

Lehman offered to sell his concession,[95] in July 1926 Guillermo Elizondo offered for sale 1,097 shares in the Apure-Venezuelan Petroleum Exploration Co. for Bs.25,000,[96] and in 1930 the Venezuelan Gulf Oil Co. proposed a secret transaction with Gómez for several concessions.[97] Although these particular offers were rejected, he nevertheless had extensive dealings in oil.

In July 1924, following the British Controlled Oilfields Ltd settlement of 31 May 1924, another British company, the North Venezuelan Petroleum Co. Ltd, which held the Jiménez Arraiz concession of 1907, and which had been threatened on and off with the annulment of its concessions, requested from the government a similar arrangement.[98] Three years later, on 23 February 1907, such an agreement was reached.[99] It contained a secret clause whereby the company agreed to transfer some of its shares to Gómez. The true amount is unknown, but on 13 September Gómez appointed Lorenzo Mercado, who represented the company in Venezuela, to manage his shares in the company.[100] Later, Gómez received shares in the Tocuyo Oilfields of Venezuela Ltd, a subsidiary of the North Venezuelan Petroleum Co. Ltd.[101] Gómez also received $450,000 and shares in the West India Oil Co., a subsidiary of Exxon, as his share of the proceeds from the concessions which Dr Márquez Rivero had acquired and transferred to the company.[102]

COMPAÑIA VENEZOLANA DE PETROLEO S.A.

All these ventures were small compared to the creation in 1923 of the Compañía Venezolana de Petróleo S.A. (CVP). In 1918 Gómez had instructed Torres to include a clause in the Oil Regulations of that year which created the National Reserves, and which was incorporated in all subsequent oil laws. It was fully acknowledged by Torres that this was a masterstroke by Gómez for it assured the Treasury a further source of revenue from the oil industry without the government having to grant further concessions, since as oil was discovered in large quantities in the adjacent operating blocks the National Reserves would naturally increase in value, and would be leased to the highest bidder.[103] However, by 1923 Gómez had decided that he should be the beneficiary of this increase in economic rent. He therefore decided to form a company that would acquire almost all the National Reserves and then negotiate them with whichever company offered the best terms. In addition, he would be able to centralize his oil activities through the company, which would handle all requests received for oil concessions, and also manage his own business deals in oil. Consequently, he entrusted his financial adviser, Colonel Roberto Ramírez, a Colombian, to promote a Venezuelan oil company. In June 1923 CVP was registered in Caracas with the object of acquiring 'concessions for the exploration and exploitation or both of oil and similar substances, to transfer them, to lease them, and to establish companies for the production, manufacture, refinement, transport and marketing of the minerals produced'.[104] It was quickly realized, however, that the company, as Willis C. Cook in July 1923,[105] and A.P. Bennett in April 1924 reported, had been formed 'with a view to enable the President and his Satellites to benefit directly and decorously by the sale of concessions'.[106] The company's prospectus, nevertheless, gave as its authorized capital Bs.25 million of which Bs.5 million had been subscribed in 1,000 shares of Bs.5,000 each. These shares were divided among the two directors, Roberto Ramírez (400), and Rafael González Rincones (300), and the company's technical adviser, Lucio Baldó (300).[107] However, none of these shareholders had actually paid for their shares or contributed towards the company's capital; both Baldó and González Rincones acted as shareholders to fulfil the requirements of the Código de Comercio.[108] It was Ramírez, however, who, as Managing Director, after an initial advance of Bs.61,000[109] from the Ministry of Development, met from his own pocket the major expenses of the installation and acquisition of concessions.[110]

The company was soon achieving Gómez's intentions: in December 1923 Rafael Falcón, an Army Commander in Zulia, received a sum of money from the company which he used to build a house,[111] in June 1924 G. Willet suggested that the company could obtain a number of concessions and then transfer them to him;[112] and in December 1925 General López Contreras was given Bs.30,000.[113] According to Lieuwen,[114] the company possessed a virtual monopoly on the National Reserves awarded by the government between 1924 and 1929,

Table 13 *Distribution of National Reserves, 1924–29*

Name	Number of National Reserves	%
CVP	177	48.1
C.A. Petróleo de Paraguaná	34	9.2
SOC (Venezuela)	28	7.6
Guillermo Elizondo	27	7.3
Others:		
companies	32	8.7
individuals	70	19.0
TOTAL	368	99.9

Source: calculated from MinFo, *Memoria,* 1924–9

though, as Table 13 shows, CVP only accounted for less than half the National Reserves, while other companies held slightly over a quarter. It should be noted, however, that many of the National Reserves awarded to other parties (as, for example, in the case of General Arturo Uzlar) were on land which was known to be of little value; furthermore, there is the possibility that some of the National Reserves were awarded to *gomecista* nominees. CVP, nevertheless, controlled a large proportion of the reserves awarded, with the added attraction that most were situated on known oil-bearing land.

In early 1924 CVP opened a New York office, headed by Edwin B. Hopkins (who also represented the company in London and Europe), with the purpose of making the acquisition of National Reserves easier. Initially it was unsuccessful because it was thought that the company's legal position was dubious, and that the concessions acquired from the company would be disputed and considered illegal upon Gómez's death. But by early January 1924 the company had begun to negotiate a massive deal with the German Stinnes Group.

In January 1924 Guillermo Sturrup and Wilhelm Waltking, representing the Stinnes Group, left Germany for Venezuela, and on 21 February Waltking signed a contract with Ramírez[115] whereby a 40-day purchase option over 200,000 hectares in

Zulia belonging to CVP was secured.[116] Moreover, the new combine would 'acquire all federal lands, or lands which became federal reserves, on more or less the same terms',[117] placing it in a very 'powerful and strategic position'.[118] The American oil companies in Caracas received the news of the contract with utter amazement since the terms of the contract appeared too onerous for the Germans. Nevertheless, both the American companies and the State Department were surprised at the magnitude of the German deal, and became increasingly apprehensive about this development for it appeared that the Stinnes Group's contract foreshadowed the creation of a strong monopoly with official backing, which by its very nature infringed upon the rights of the U.S. companies. Consequently, the companies, backed by the U.S. Legation, at first threatened to contest the legality of the Stinnes Group acquiring National Reserves from CVP.[119]

The Venezuelan government did not take this seriously, and Gómez informed Chabot, the American Chargé d'Affaires, through Rafael Requena, that the 'American petroleum interests in Venezuela need not fear anything as he will permit no one to take advantage of them'.[120] The impression which government officials wanted to create was that CVP was a strictly private company, and 'one whose activities can not be of legitimate interests to foreigners, and particularly to the Government of the United States'.[121] It was also felt that the Stinnes Group's contract would force the large companies to exploit their concessions, which government officials suspected were being kept by these companies as reserves for the future. But of more importance still to Gómez's financial interests was the belief that the large companies had acquired all the land they needed and that they would make no further purchases in the foreseeable future. It was therefore necessary to find new purchasers.[122] But the position of the American companies was weakened when it was known that R. Bingham and Colonel Armstrong of Exxon had decided that CVP's titles were valid as long as all the legal requirements had been met and had been approved by Congress,[123] and that they had therefore decided to negotiate several oil titles from CVP.

Much to the relief of the oil companies, the Stinnes Group, through financial difficulties, was unable to take up its

option.[124] At the same time there was a growing feeling that the
Stinnes Group affair had been orchestrated by the government
in order to 'stimulate activity in other petroleum interests in
Venezuela'.[125] Although this view cannot be dismissed
altogether, the assertion appears highly unlikely because
Gómez's objective in agreeing to the Stinnes Group's contract
was twofold: on the one hand, he wanted to benefit financially,
and on the other, he wanted to create a strong third party
which would compete with the established Anglo-Dutch interests and the growing American presence, and which would
lessen the country's dependence on the Anglo-American oil
companies. Moreover, as has been pointed out, at least one
major company was willing to acquire National Reserves from
CVP before the Stinnes Group withdrew.

But the companies resented the unfair treatment afforded
them in the acquisition of National Reserves, and although
Chabot, and the adverse publicity in the U.S., forced the
government to publish in the Official Gazette details about
when the competitive bidding for National Reserves would take
place, the time span between this and the awarding of the
contract to CVP was so short that 'no Company outside the
National Petroleum Company has had the opportunity to
investigate the concessions or bid on them'.[126] However, the
general disagreement among the companies in handling the
question of CVP and the National Reserves worked in CVP's
favour because once one company had decided to do business
with it the other companies would of necessity follow suit. It
thus became apparent to the oil companies that the government
would treat National Reserves as Gómez's private domain,[127]
something which during the ensuing months the companies
were forcefully reminded of when all the National Reserves were
awarded either to CVP or to individuals closely connected to
Gómez. The companies, therefore, reluctantly accepted the role
of CVP. It also became apparent that any company which
entered into an agreement similar to the Stinnes Group's
contract would be in a strong position. Shell, for example, tried
to secure such an agreement in late 1924, but the terms offered
were unacceptable to CVP,[128] which proposed instead to
develop 30,000 to 50,000 hectares it owned on a partnership
basis with Shell. Although these negotiations were not resolved,

CVP was more successful with other companies. For example, it sold 15,000 hectares to the Venokla Oil Co. for $250,000 (Bs.1,300,000) and a 2.5 per cent production royalty, and 83,000 hectares to the American–Venezuelan Oilfields C.A. (a subsidiary of Buckley's Pantepec Oil Co. of Venezuela) for $750,000 (Bs.3,900,000).[129] In addition, the following companies took out purchase options with CVP: the International Petroleum Co., 30,000 hectares for $350,000 (Bs.1,820,000) and a 2.5 per cent production royalty; the Esperanza Co., 225,000 hectares for $350,000 (Bs.1,820,000) and a 2.5 per cent production royalty; and the Venezuelan Gulf Oil Co., 10,000 hectares for $110,000 (Bs.570,000) and a 2.5 per cent production royalty.[130] In June 1926 CVP negotiated a further 120,000 hectares with the Venezuelan National Petroleum Corp. for $1,650,000 (Bs.8,580,000) and a 2.5 per cent production royalty.[131]

In most cases CVP received a production royalty on the land transferred, thus securing a small but important stake in the future prosperity of the operating companies. But it was also apparent that its own future prosperity would be better served if it entered into a joint developing partnership, as had been attempted, albeit unsuccessfully, with Shell, with various oil companies. Diógenes Escalante, the Venezuelan Minister in London, convinced Gil Fortoul in early 1926 of the need for CVP to enter into such a partnership with other oil companies, for he argued that the development of the British oil companies in Venezuela demonstrated 'the economic mistake which we made disposing of the concessions for absurdly low cash settlements'[132] instead of participating with them in a joint development company.[133] He further advised Gómez to offer the National Reserves on an equal basis to Anglo-American capital, as in this way the Venezuelan government would have greater control over the parent company, and as this would also curtail the monopolizing effect that Shell and Exxon had on the oil industry.[134] Gómez, who had not been privy to the Shell talks, took up this suggestion and sent Gil Fortoul to negotiate the National Reserves held by CVP on a partnership basis. With the help of P.C. Heyden and Antonio Aranguren, Escalante and Gil Fortoul made rapid progress in getting a company interested in the oil titles, as well as in several mining titles for

Guayana which Gil Fortoul was also negotiating.[135] In November 1926 Gil Fortoul reached a tentative agreement with the South American Oil and Development Corp., an American company registered in Delaware on 31 March 1925,[136] and returned to Venezuela. After discussing the matter thoroughly with Gómez, he was back in London in June 1927 to finalize the British arrangements, and subsequently travelled to Paris to make the final arrangements and to sign the contract, which he did on 15 October 1927. Under the agreement Gil Fortoul promised to transfer to Clinton D. Winant of the South American Oil and Development Corp. all the concessions held by CVP (1,586,062.9 hectares in exploration titles, and 210,585.68 hectares in exploitation titles), and a few more to be obtained from the Venezuelan government. The company had three months in which to decide whether to take up its option on National Reserves, and, if successful, a new company would be formed within six months with an authorized capital of £2 million subscribed by British, American, and French interests.[137] Although the transfer document states that CVP received a cash settlement for its concessions, a private agreement[138] was reached by which Gil Fortoul would receive 50 per cent of the net profits of the company, and which Gómez distributed thus:[139] 10 per cent to the government, 15 per cent to Gil Fortoul, and 25 per cent to CVP.[140]

As soon as the details of the contract were made known to CVP's Board of Directors, they sent an extensive Memorandum to Gómez detailing the disadvantages to his financial interests with the contract entailed. Under the contract the concessions acquired from CVP were to be transferred after all the taxes had been paid, which amounted to Bs.782,000; furthermore, the company could return the concessions at the end of its six-month option and then CVP would have to pay taxes during these months of Bs.210,000 as well as the transfer and registry fees, amounting to a further Bs.200,000. This meant that if the company decided not to purchase the options, CVP would have to pay Bs.1,192,00 in taxes, of which the company was only obliged to return Bs.375,000, 'although payment was not guaranteed by the contract'.[141] Thus, Gómez would have paid out a minimum of Bs.817,000 without any profit, and in addition the company's rejection of the land would mean that

CVP's titles would be devalued since potential future buyers would view such rejected land with disfavour.[142] But even if the company did take up its option, Gómez would be worse off because it would have acquired Bs.26,318,142-worth of oil lands in return for 40 per cent[143] of the net profits, which was subject to the discovery of oil and to the development of the property. For Ramírez *et al.*, the company's intention appeared to be speculative since its authorized capital of £2 million had not been subscribed. It could therefore retain the lands and transfer them later to other companies, and/or make speculative gains on the Stock Exchange. Nevertheless, if the company developed its property, the likelihood of Gómez receiving his share of the profits in the near future was remote because of all the initial expenses incurred by the company.[144] Moreover, in addition to the taxes already mentioned, Gómez's profits would be subject to taxes in the other countries where the company operated. Finally, the company reserved the right to an annual sum free of tax, equivalent to 10 per cent of the company's paid-up capital, which would come out of Gómez's share of the profits.[145]

Despite this, on 13 December Alamo informed CVP that Gómez had given him permission to authorize them to 'transfer to Dr. Gil Fortoul, or to the person or company which he indicated'[146] all concessions discussed above. Two days later, on 15 December, this was effected. At the same time, on 22 December, the South American Oil and Development Corp. was registered in Caracas, and a subsidiary, the Venezuelan International Petroleum Corp., was formed to hold all the concessions acquired from Gil Fortoul. But Naphen, who managed the South American Oil and Development Corp., had started negotiations with Exxon to transfer to it the South American Oil and Development Corp. and the Creole Syndicate. As a result, at an Extraordinary General Meeting of the Creole Syndicate shareholders, it was agreed to change the company's name to Creole Petroleum Corp., and to increase its authorized capital from 2.5 million to 6 million shares of no par value, in addition to issuing 500,000 shares to acquire the stock of the South American Oil and Development Corp.[147] Once this took place, Exxon acquired 3,025,000 shares of the Creole Petroleum Corp. and made it a holding company,[148] transferring to it all the concessions, equipment, and buildings held by its subsidiaries in Venezuela, which included the former concessions held by CVP.

Despite these developments, by mid-June 1928 Gómez was taking a greater interest in CVP's initial objections to Gil Fortoul's contract.[149] Gil Fortoul was able to reassure Gómez that his shares in CVP were not worthless as a consequence of the deal, and that the Venezuelan International Petroleum Corp. would, contrary to suggestions made by critics of the deal, declare a dividend long before ten years had elapsed. As proof of this, Gil Fortoul was authorized to propose the following three schemes to Gómez: first, to acquire Gómez's stock in CVP for Bs.8 million;[150] second, to set up a refinery in Venezuela which would utilize the 10 per cent royalty which the government received, with two parties sharing profits; and third, to negotiate any further concessions acquired by CVP.[151] Gil Fortoul returned to Paris and transferred the government's 10 per cent production royalty to the General Oil Development Co., headed by Edward Herbert Keeling. Gómez was still unsure, but Gil Fortoul urged him to accept because 'according to my calculations the Government would receive many more millions than it does today, if we reach an acceptable and advantageous agreement both for the government and the capitalists involved in the business, and a large profit margin will remain to be distributed among the people whom you indicate'.[152] Consequently, a month after Gil Fortoul's return to Caracas in October 1928, a contract was signed with Keeling.[153]

However, the contract between CVP and the South American Oil and Development Corp. appeared in 1929 to be heading for trouble when Exxon discovered that CVP had not paid the taxes on many of the titles transferred to the South American Oil and Development Corp. The contract had clearly stipulated that the concessions belonging to CVP 'should be free of any taxes on the date of transfer',[154] and was signed and approved by Alamo, but CVP had not paid the outstanding taxes because Alamo had assured the company that the taxes would be cancelled.[155] In early 1929 Gil Fortoul entered into an agreement with the company by which he consented to pay Bs.946,329[156] in outstanding taxes, while the company would pay the registry fees of Bs.180,000.[157] But Gil Fortoul had cancelled only the taxes owed on the concessions which he had transferred to the company in October 1927, and not those owed for the six months during which the company had held its purchase option. Consequently, Exxon found itself in the position of either paying the oustanding taxes or losing the concessions by default. It was forced to request Gil Fortoul and CVP to settle the debt, and also to transfer those concessions included in the contract which were pending free of any tax debts, something CVP could not do because of a lack of funds.[158]

The tax problem was having a serious effect on the exploitation of the concessions and consequently on the financial benefits accruing to Gómez. As a way out of the impasse, Gil Fortoul suggested that the Government write off the taxes or that CVP pay the taxes owed.[159] Ramírez countered by proposing that the deal be renegotiated with Exxon and that Gil Fortoul's 15 per cent share in the profits be transferred to CVP to allow it 'complete freedom of action'.[160] However, no agreement was reached, and in January 1930 Exxon decided to transfer several of the unpromising concessions which it had acquired from Gil Fortoul to the South American Oil and Development Corp. on the understanding 'that it would re-acquire them as soon as the contractual obligations of the Compañia Venezolana de Petróleo and the Parish agreement had been complied with in favour of the South American Oil and Development Corporation'.[161]

Nevertheless, the stalemate continued and by 1931 the South American Oil and Development Corp. owed the Treasury Bs.2 million in back taxes, a sum which was impossible for CVP to settle because it could not transfer the concessions as no company was willing to take on such a tax burden.[162] If the concessions were renounced and then reacquired by CVP the company could sell them and pay the Treasury, but if the concessions were allowed to expire the other companies would acquire them, thus depriving CVP of the only possible way of paying the government.[163] Moreover, as Liscano suggested in June 1931, another oil company could be organized with new titles and this company would then pay back the outstanding taxes to the government.[164] Although it is uncertain what occurred after this, it is likely that the problem with Exxon was resolved amicably.

Gómez's dealings with National Reserves was not confined only to CVP; for example, in 1928 he promised José Ignacio Cárdenas, then the Minister of Public Works, several National Reserves in Monagas State,[165] and in 1929 he requested Torres not to consider any offer for the Venezuelan Gulf Oil Co.'s National Reserves because he would indicate in due course 'the person to whom it should be transferred, for the purpose of a deal which is being negotiated with the said company'.[166]

The creation of CVP served Gómez well for he was able to reap considerable financial benefits. The Jurado de Responsabilidad Civil y Administrativa estimated in 1946 that Gómez's share of CVP's profits was Bs.20,641,086,[167] but this seems far too conservative an estimate for it did not include the secret agreement with Exxon's subsidiary which gave him a quarter of the profits on the concessions transferred (and part of the government's 10 per cent profit share), and also assigned to him a 2.5 per cent production royalty for the duration of the contract. Nor did it take into account the 2.5 per cent production royalty stipulated by most contracts entered into by CVP prior to Gil Fortoul's 1927 deal, in addition to the normal cash settlements. These same royalties linked Gómez and his heirs directly with the oil industry, thus creating a powerful vested interest which would stimulate the industry's growth. The same interest was also extended to other concessionaires and/or intermediaries who held similar production royalties, thus establishing a strong and influential local lobby which the companies could call on if changes needed to be made to their operating conditions. Table 14 shows the number of concessionaires under the principal oil holding companies.

As it has been shown, there was a clear vested interest on the part of Gómez, his family, entourage, and government officials (as well as individual concessionaires) to stimulate the development of the oil industry. A thriving industry would facilitate the transfer of concessions at higher prices, and, with many settlements negotiated on the basis of a fixed amount per barrel, higher production volumes meant faster returns, as well as benefiting the large number of concessionaires who had negotiated production royalties. Consequently, there was a strong and

powerful vested interest which identified itself firmly with the
oil industry.

Table 14   *Number of concessionaires under the principal oil holding
companies 1907–36*

Name of holding company	Number of concessionaires	%
Exxon	187	22.6
Alfred Meyer	84	10.1
Sinclair Oil Co.	82	9.9
Gulf Oil	73	8.8
Pantepec Oil Co. of Venezuela	44	5.3
Sun Oil Co.	28	3.4
SOC (California)	25	3.0
Andes Petroleum Co.	24	2.9
Atlantic Refining Co.	20	2.4
Brokaw, Dixon and Gardner	20	2.4
Others (61)	242	29.2
TOTAL	829	

Note: most concessions transferred in the early 1920s did not include a
production royalty, but as the industry grew most transfer contracts stipulated
such a royalty
Source: ADCOTHMEM, 'Historial de Concesiones de Hidrocarburos', Vols.
1–34, Files 1–16,620. For the full distribution see Appendix A

4
National and local effects of the oil industry

The direct impact of the oil industry was felt first in Zulia State, and only later in the rest of the country. Zulia's socio-economic structure was severely disrupted by the oil boom because it created a scarcity of labour for the traditional agricultural sector, and because it had a tremendous inflationary effect on the local economy. The oil boom, however, stimulated the expansion of the commercial elite of the country which during the 1920s and 1930s strengthened its position within the economy to the detriment of the agricultural elite. Although the period was one in which agriculture was in decline, it should be noted that up to 1936 it contributed substantially to the country's economic growth:[1] for example, according to Rangel, Táchira State's agricultural production and exports (coffee and cattle) continued to rise during the 1920s.[2] Nevertheless, the disparity between the oil and agricultural sectors widened during this period.

Venezuela's total trade during Gómez's regime increased enormously: for example, between 1908 and 1929 exports increased from Bs.83 million to Bs.739 million, a 790 per cent increase due mainly to the oil boom of the 1920s.[3] Oil exports during the mid-1920s displaced traditional exports of coffee and cocoa. The traditional markets for Venezuelan exports were the U.S., France, and Germany, while Britain was the largest single supplier of foreign goods. During the First World War, when European markets were closed and there was a lack of adequate transport facilities, the U.S. increased its trade with the whole of South America. In Venezuela, for example, American exports increased from 37 per cent to 64 per cent between 1914 and 1916, and Venezuelan exports to the U.S. increased by 25 per cent during the same period.[4] The war stimulated economic

109

activity in the country because the high world prices of manu-
factured goods allowed many nascent industries in Venezuela to
gain ground. In addition,the high world prices of agricultural
products stimulated exports of sugar, corn, lard, etc., together
with the more traditional exports.[5] Consequently, in 1920 the
Venezuelan government appointed a number of Special Com-
missioners in Europe and Japan to stimulate and increase trade
between these countries and Venezuela. Nevertheless, the U.S.
retained and increased its share of Venezuela's market during
the 1920s. For example, the value of Venezuelan exports in 1926
to America was Bs.229 million compared to Bs.53 million, Bs.38
million and Bs.26 million to Britain, Germany, and France
respectively.[6] The American presence was further strengthened
during 1927 when a permanent office of the Foreign Commerce
Services of the U.S. government was established in Caracas.[7]
Nevertheless, Venezuela, despite its booming oil economy, was
a relatively unimportant trading partner of the U.S., maintaining between 1929 and 1936 only a 5–6 per cent share of the
total imports and exports into the U.S. from Latin America.[8]
    Trade was very important to the Venezuelan government
since it derived its most important source of revenue from
custom duties levied on imported products.[9] Consequently, the
government welcomed any economic boom which would lead to
an increase in trade. In order to reduce the burden of customs
duties on imported goods, the government in 1914 increased the
tax on salt, cigarettes, and liquor. Oil revenues, which started
flowing during the mid-1920s, provided the government with a
new source of revenue free from the vicissitudes of other trade
(during 1921–2 trade had decreased considerably, causing a
decline in government revenues). The government would still,
however, depend heavily on customs receipts, for although oil
revenues at the end of Gómez's regime accounted for nearly a
third of its total revenues,[10] customs duties were still its most
important source of finance (see Table 15).[11]
    In December 1908 Gómez inherited from Castro a budget
deficit of Bs.5,108,531, and for the next two years the government's budget was in the red, but from then onwards Gómez
was able to maintain fiscal control. Government revenues
increased from Bs.50.4 million in 1908–9 to Bs.225.5 million in
1929–30, decreasing to Bs.188.9 million in 1931, and slightly

Table 15 *Percentage distribution of government revenue, 1908–29*

Year	Customs receipts	Tobacco, Liquor and Salt tax	Stamp tax	Mines	Others
1908	70	20	5	0.2	4.8
1909	70	13	10	0.4	6.6
1910	72	11	9	0.3	7.7
1911	70	9	8	0.6	12.4
1912	76	17	2	0.6	4.2
1913	72	17	2	0.8	8.2
1914	59	29	5	0.4	6.6
1915	60	28	6	0.6	5.4
1916	60	27	6	1.0	6.0
1917	44	39	8	0.8	8.2
1918	40	41	9	3.0	7.0
1919	51	26	7	2.0	3.0
1920	52	28	9	2.0	9.0
1921	38	33	9	3.0	17.0
1922	47	29	8	9.0	8.0
1923	45	26	7	5.0	17.0
1924	55	24	7	7.0	7.0
1925	51	18	6	15.0	10.0
1926	56	18	7	11.0	8.0
1927	50	19	7	19.0	5.0
1928	49	15	6	20.0	10.0
1929	51	13	6	21.0	9.0

Note: fiscal years run from 1 July to 30 June
Source: FO 199/266, C.N. Clark, 'A survey of the income, expenditure, Treasury balance, and status of the funded debt of Venezuela, annually, during General Gómez's direction of the Government', July 1929; and U.S. Department of Trade, Charles J. Dean, 'Commerce and industrial development in Venezuela', *Trade Information Bulletin*, No. 783 (1931), p. 12

increasing to Bs.206.4 million at the end of Gómez's regime.[12] Gómez was very careful to balance his budgets, and in the period 1908–31 achieved a surplus of Bs.40.2 million, which allowed him to ease the country's internal and external debt.

Government expenditure remained remarkably stable in certain areas, notably for Internal Affairs, Foreign Affairs and Education. During the years in question there was a shift away from the War and Navy Ministry (especially after 1913 when Gómez consolidated his power), and from the Treasury, to the Ministry of Public Works, which increased its budget from 7 per cent in 1908 to just under 33 per cent during the latter half of the 1920s (see Table 16).

Gómez's greatest financial achievement was in paying off the country's large foreign debt, and in substantially reducing the internal public debt. In January 1909 the foreign debt stood at Bs.225.5 million, by 1918 it had been reduced by 36 per cent to Bs.143.5 million,[13] being paid off in full on 17 December 1930 in commemoration of the centenary of Bolívar's death.[14] The internal debt in 1908 stood at Bs.67.5 million, and was reduced to Bs.28.7 million in 1929, and almost completely paid off by the time of Gómez's death. The foreign debt was repaid much faster than the internal debt, so that by the mid-1920s, when the oil revenues assumed importance, the country's debt had been considerably reduced. Nonetheless, oil revenues contributed to an acceleration in the rate of repayment.

The oil industry, then, had a positive effect on government revenues, significantly cushioning the country during the late 1920s and early 1930s from the effects of the Great Depression, and in so doing providing for the country's financial and political stability. Nevertheless, there were other areas of economic activity equally affected by the industry, either directly or indirectly, which did not fare so well, and which need detailed examination.

THE ECONOMIC IMPACT OF THE OIL INDUSTRY

The oil industry, between 1925 and 1936, grew at an average annual rate of 75 per cent, whereas agriculture lagged behind at 3.5 per cent, industry at 11.3 per cent, and commerce and services at 9.9 per cent annually (see Table 17). The oil

Table 16 *Percentage distribution of government expenditure, 1908–27*

Year	Interior	Foreign Affairs	Treasury	War and Navy	Development	Public Works	Education
1908	16	8	40	19	6	7	4
1909	23	2	41	18	7	6	3
1910	24	2	42	14	5	5	8
1911	29	1	37	17	5	6	5
1912	26	1	25	20	6	16	6
1913	24	2	22	32	5	8	6
1914	30	1	23	25	6	8	6
1915	27	1	26	18	5	19	4
1916	29		25	20	6	14	5
1917	28	2	25	20	7	13	5
1918	30	2	27	16	8	12	5
1919	28	2	22	16	9	19	4
1920	21	6	18	20	8	23	4
1921	28	4	18	18	10	16	6
1922	27	5	22	18	9	13	6
1923	27	4	22	16	9	16	6
1924	32	3	14	13	7	27	4
1925	33	2	13	10	6	33	3
1926	27	2	11	9	16	32	3
1927	30	3	13	11	11	28	4

Government Ministries

Source: calculated from FO 199/266, C.N. Clark, 'A survey of the income, expenditure, Treasury balance, and status of the funded debt of Venezuela, annually, during General Gómez's direction of the Government', July 1929

113

Table 17 *Gross National Product, 1925–36 (at 1936 prices)* *(in million Bs.)*

Year	Total	Agriculture	%	Petroleum	%	Crafts and industry	%	Commerce and services	%
1925	726.7	252.0	34.6	69.1	9.5	139.6	19.2	266.0	36.6
1926	755.8	243.8	32.3	110.7	14.6	135.1	17.9	266.2	35.2
1927	800.4	259.0	32.3	141.1	17.6	140.9	17.6	259.4	32.4
1928	977.1	255.4	26.1	237.8	24.3	164.4	16.8	319.5	32.7
1929	1188.5	283.6	23.9	341.1	28.7	193.5	16.3	370.3	31.2
1930	1289.1	276.1	21.4	419.1	32.5	207.6	16.1	386.3	30.0
1931	1074.6	291.8	27.2	386.2	35.9	173.1	16.1	223.5	20.8
1932	1201.7	297.1	24.7	413.2	34.4	175.6	14.6	315.8	26.3
1933	1298.5	295.4	22.7	488.1	37.6	185.2	14.3	329.8	25.4
1934	1482.6	312.7	21.1	575.0	38.8	214.8	14.5	380.1	25.6
1935	1746.1	328.7	18.8	651.0	37.3	274.6	15.7	491.8	28.2
1936	1865.8	350.1	18.8	646.6	34.7	312.9	16.8	556.2	29.7

Source: adapted from Domingo Alberto Rangel, *Capital y Desarrollo. El Rey Petróleo* (2 vols., Caracas, UCV, 1970), Table 21

industry, which dominated the economy in the latter half of Gómez's regime, increased considerably the per capita income of the country which, according to Córdoba, rose by 45 per cent between 1920 and 1928, while the national income increased by 74 per cent.[15] This had a relatively small inflationary effect on prices in general in the country, and prevented prices from dropping during the mini world recession of 1920–2. According to Acedo De Sucre and Nones Mendoza, prices between 1913 and 1928 increased by 28 per cent, but then up to 1935 decreased, as the recessionary effect of the Great Depression was felt, to 23 per cent below the prices prevailing in 1913.[16]

The recession of the early 1920s hit Venezuelan traders very hard, but by the mid-1920s the crisis appeared to be over, due to the injection of oil money. For example, in San Cristobal (Táchira State) it was reported that houses which in 1922 had been leased for Bs.80 per month were fetching Bs.200 per month in 1926; and that eggs, which in 1922 had cost Bs.0.50 for four, had doubled in price four years later.[17] The sudden influx of oil money and the high price of coffee in the mid-1920s combined to create the impression that the boom would continue forever, with the result that many overtraded and overspent, much of the latter going on imported luxuries.[18] When the oil companies cut back their operations in 1927 and coffee prices declined, many trading firms were unable to cancel their debts or meet their foreign commitments.[19] But given the importance of the oil industry and its dominance of the Venezuelan economy, it might have been expected to have had a much greater inflationary impact on the economy. This did not occur because a large part of the industry's national product was repatriated as profit income. Consequently, the direct impact the industry had on the economy must be examined. This can be measured by the 'retained value' of oil, which is the part of the value of output to the industry retained in the country in the form of tax and other payments to the government, wage payments and purchases of domestic goods, and non-wage services. Table 18 shows that the direct domestic impact of the oil industry between 1925 and 1935 was on average 11.0 per cent of the GNP. Although this is significant it is on average slightly over a third of the impact attributed to it if the industry's percentage share of the GNP is considered.

Table 18 *Direct domestic impact of the petroleum industry (in million Bs.) as a percentage of Gross National Product, 1925–35 (in million Bs.)*

Year	Total government payments	Cash shares	Domestic wages	Domestic purchases	Retained value	GNP	Retained value as % of GNP
1925	21	17	36	20	94	726.7	12.9
1926	18	19	48	20	105	755.8	13.9
1927	21	19	62	20	122	800.4	15.2
1928	46	29	62	20	157	977.1	16.1
1929	51	14	80	20	165	1,188.5	13.9
1930	47	18	67	20	152	1,289.1	11.8
1931	47	5	38	20	110	1,074.6	10.2
1932	45	1	28	20	94	1,201.7	7.8
1933	45	5	34	20	104	1,298.5	8.0
1934	52	3	46	20	121	1,482.6	8.2
1935	59	21	50	20	150	1,746.1	8.6
1925–35	452	151	551	220	1,374	12,541.1	11.0

Source: calculated from Manuel R. Egaña, *Tres décadas de producción petrolera* (Caracas, Tip. Americana, 1947), Table 12, p. 15; Ramón David León, *De agro-pecuario a petrolero* (Caracas, NP, 1944), pp. 68–70; and Domingo Alberto Rangel, *Capital y Desarollo. El Rey Petróleo* (2 vols., Caracas, UCV, 1970), Tables 21 and 35

This increase in effective demand led to increased levels of consumption, especially of imported goods,[20] while weakening local manufacturing industry and increasing the commercial importance of Caracas. According to Stann, between 1891 and 1936 manufacturing firms in Caracas declined from 34 per cent in 1891 to 24 per cent in 1936, while the number of commercial establishments increased from 58 per cent in 1891 to 68 per cent in 1936,[21] with half the capital invested in this sector concentrated in banks, wholesale dry foods and wholesale foodstuffs. The structure of manufacturing industry between 1891 and 1936 remained the same except for an increase in the food and clothing industries to keep pace with the growing population. The oil industry, it is apparent, had very little effect on the development of the economic base of the country, and did not stimulate investment in the capital goods sector of the economy.[22]

It should be noted that the relatively small impact of the oil industry on the country's economy was a function of the size of the industry and not due to excessive oil revenues being repatriated out of the country. Table 19 shows government oil revenues between 1922 and 1935 to have been 8.5 per cent of the value of oil exported during that period. However, if we measure the retained value of oil in Venezuela as a percentage of the total value of oil, we find that for the same period nearly 28 per cent of the oil's value was retained in the country. This relatively large figure is confirmed by a study by Claudio Urrutía in 1934 of the benefits accruing to Venezuela from the oil industry, in which he found that 37 per cent of the total value of oil shipped from Venezuela was retained in the country.[23] This was considerably less than in Mexico where from 1926 $2 were retained for every $3 received for its oil.[24]

AGRICULTURE

The effect of the oil industry on agriculture was of crucial importance since the vast majority of the country's population was gainfuly employed in this sector. Coffee was the country's largest export product before the advent of the oil industry, and up to 1913 (when Colombia overtook it) Venezuela was the second largest coffee producer in the world after Brazil. Venezuela's share of the world market between 1909 and 1939

Table 19   *Government revenues and retained value of oil as a percentage of the value of oil, 1922–35 (in million Bs.)*

Year	Value of oil	Government revenues	%	Retained value of oil	Retained value of oil as % of the value of oil
1922	15	7.5	50.0	41	273
1923	33	3.8	11.5	45	136
1924	82	5.9	7.2	67	82
1925	186	20.9	11.2	94	51
1926	334	17.9	5.4	105	31
1927	338	21.4	6.3	122	36
1928	564	46.2	8.2	157	28
1929	829	50.5	6.1	165	20
1930	864	47.3	5.5	152	18
1931	568	47.0	8.3	110	19
1932	612	45.1	7.4	94	15
1933	284	44.8	15.8	104	37
1934	380	52.0	13.7	121	32
1935	428	59.3	13.9	150	35
1922–35	5,517	469.6	8.5	1,527	27.7

Source: calculated from Alirio Parra, *La industria petrolera y sus obligaciones fiscales en Venezuela* (Caracas, Primer Congreso Venezolano del Petróleo, 1962), Table D-6; Manuel R. Egaña, *Tres décadas de producción petrolera* (Caracas, Tip. Americana, 1947), Table 12, p. 15; Ramón David León, *De agro-pecuario a petrolero* (Caracas, NP, 1944), pp. 68–70; and Domingo Alberto Rangel, *Capital y Desarollo. El Rey Petróleo* (2 vols., Caracas, UCV, 1970), Tables 21 and 35

declined from 4.5 per cent to 2.8 per cent, while Colombian production increased its market share from just under 5 per cent to 10 per cent during the same period.[25] Nevertheless, up to Gómez's death, coffee still remained the country's largest cash crop and agricultural export. The U.S. was the single largest market for Venezuelan coffee, but the European countries of Germany, France, Spain and Holland accounted for over 60 per cent of the country's exports.[26] Since the mid-1920s, however, oil exports completely overshadowed Venezuela's other exports (see Table 20).

Table 20   *Percentage distribution of Venezuela's principal exports*
by value, 1925–36

Year	Coffee	Cocoa	Petroleum	others
1925	38.1	9.0	41.7	11.3
1926	25.0	4.8	61.5	8.7
1927	23.3	6.1	63.5	7.1
1928	13.7	4.4	76.6	5.3
1929	17.2	3.1	76.3	3.5
1930	8.9	2.3	87.3	1.5
1931	10.0	2.1	84.1	3.8
1932	9.3	1.9	84.6	4.2
1933	5.4	1.5	89.6	3.5
1934	4.9	0.9	90.6	3.6
1935	4.3	0.9	91.2	3.5
1936	5.2	1.5	89.0	4.3

Source: adapted from *Revista del Instituto Nacional de Café*, 1:3 (May 1940)

Although coffee exports decreased during Gómez's period, total coffee production remained stable. According to Garaicoechea agricultural production as a whole increased between 1920 and 1936 by 2–6 per cent. This modest increase was insufficient to supply the growing demand created by population growth and increased per capita income resulting from the oil boom, which meant that less goods, such as coffee and cattle were available for export. This inadequacy of the agricultural sector to meet growing demand was almost unanimously blamed on the pernicious effects the oil companies had on the supply of labour and on the supply of good agricultural land. Adriani, for instance, argues that the demand for labour created by the oil companies induced the relatively small number of rural workers to flock to the oilfields.[27] But the problem was more complex than this; for example, world-wide coffee prices during the 1920s remained buoyant, allowing Venezuelan producers to sell their diminishing production at higher prices. However, during the 1930s coffee prices dropped considerably, making Venezuelan coffee uncompetitive because of its relatively high production costs compared to other Latin American producers; and, in addition, Venezuelan coffee exporters were penalized by the adverse exchange rate which the influx of oil-dollars had on the country's currency.

Venezuela's uncompetitiveness was the result, in the main, of low levels of investment caused by the high interest rates charged on capital borrowed. Delgado Palacios in his study of coffee production in Venezuela found that labour costs (which in 1895 represented 18 per cent of total costs) were relatively small compared to costs associated with transport and interest repayments, which accounted for 40 per cent of total costs, while capital and installation costs were 29 per cent and cultivation costs 14 per cent.[28] Despite the high profits obtained by some coffee growers, especially during the boom following the First World War, high interest charges and transport costs resulted in disinvestment by producers, with a consequent decline in coffee yield per tree. For example, in 1947 it was estimated that production per tree in Venezuela was on average 126 *granos*, compared to 500 *granos* in Costa Rica and Guatemala, and just under this figure in Colombia and Brazil.[29] Consequently, production costs were much higher in Venezuela (on average Bs.35.47 per 46-kilogram sack in 1938) compared to Colombia (Bs.20.09), Nicaragua (Bs.17.6), and Guatemala (Bs.14.6).[30] But there was another important aspect that affected coffee production, which was the structure of land ownership by coffee producers. While production in Colombia increased, so did the number of smallholders producing coffee, whereas the reverse occurred in Venezuela, especially in Táchira State. The importance of this increase in the number of smallholders, according to Linares, is that coffee production on a small family farm is constant because the family does not work 'for immediate gain but for the future'.[31] On large haciendas, however, labour became scarce because it was only needed during harvest time, and earnings from such a short period could not maintain the coffee picker for the rest of the year. With small family plots labour supply is less of a problem because the whole family is involved in coffee production, and is not tempted to cease growing coffee permanently. Certainly, in areas where coffee was produced on family plots, as in the states of Sucre, Monagas, and Trujillo coffee production increased during the period in question.[32] However, a lack of a large population meant that the country's agricultural potential was not fully developed; for example, in Táchira at harvest time, Colombians would be brought in to supplement a deficient labour supply. This was further accentuated when, as a result of the oil companies' activities during the 1920s, a great number of people were attracted to the oil centres of Zulia. (It should be noted that the Andean states of Trujillo, Mérida, and Táchira in 1946 produced only 58 per cent of the country's total crop.)[33] While the total production diminished from 81.6 million kilograms in 1919 to 45.4 million kilograms in 1939, the total area under cultivation remained roughly the same over the period. Nevertheless, the agricultural sector, because it was the biggest employer of people, and because up to the early 1930s it was the largest source of earnings, was still of considerable importance to the economy of the country (see Table 21). According to Veloz Goiticoa in 1919, 24 per cent of the country's population depended directly on agriculture,[34] and the 1926 Census revealed that out of a total of 812,492 active workers, 56 per cent were in agriculture, 32 per cent in industry, and 12 per cent in commerce.[35] Consequently, Venezuela, despite her wealth derived from oil and related industries, was still an agricultural country which depended for much of the well-being of its population on the prosperity of its agricultural sector, and on its exports.

The oil recession of 1927, and the consequent release of men from the industry, once again focussed attention on the need to stimulate agriculture. The *Boletín de la Cámara de Comercio de Caracas* again urged the establishment of an agricultural credit bank, and also stressed the need for an increase in the government's programme of public works and road construction in order to stimulate demand in various parts of the country.[36] The government, encouraged by Dr Abel Santos (the former Minister of Finance),[37] was now preparing its Banco Agrícola y Pecuario Bill to be presented to Congress in 1928. Such a bank, however, was to have limited effect. With a capital of U.S. $6 million (financed wholly out of current revenue), the bank replaced the German trading houses as the principal agricultural financier. The landowners would mortgage their land to the bank, and use this money to finance mercantile or construc-

Table 21   *Comparison of income of the two largest economic sectors,*
1922–36 (in million Bs.)

Year	Petroleum	% of total	Agricultural and mining sectors	% of total
1922	22.0	14	122.1	86
1923	25.4	17	128.2	83
1924	48.0	24	148.0	76
1925	75.4	28	192.5	72
1926	87.8	37	148.4	63
1927	107.7	40	163.3	60
1928	144.3	50	142.6	50
1929	154.2	45	184.9	55
1930	152.1	54	128.4	46
1931	98.6	49	103.8	51
1932	81.4	46	96.6	54
1933	91.5	59	64.3	41
1934	105.3	62	63.4	38
1935	137.7	69	62.4	31
1936	138.4	62	84.2	38

Source: adapted from Domingo Alberto Rangel, *Capital y Desarrollo. El Rey Petróleo* (2 vols., Caracas, UCV, 1970), Table 18

tion businesses. The bank in its haste lent money with such small guarantees that when the time came to repay the loan, the landowners preferred to keep their new businesses since the value of land had dropped on account of the Great Depression.[38] Others went bankrupt because they were unable to pay back the loans.[39] The bank, therefore, did not serve to alleviate the lack of agricultural investment, which was the major problem facing agricultural workers and coffee growers. The bank, however, did affect the pattern of investment in agriculture, especially in the areas around the oil camps, where the price of land naturally increased beyond the reach of the average farmer. Writers such as Brito have argued that many oil companies became new *latifundistas* because of their power of appropriation, and that many of the appropriated landowners became bankrupt because they received a paltry rent ranging from between Bs.0.75 and Bs.3.25 per hectare per annum.[40]

But a close examination of the land held by oil companies reveals that the oil industry occupied a very small percentage of the total land surface of the country; and, in addition, the type of land occupied was very poor. According to Dupouy, oil-producing fields 'are situated in poor land . . . and also in woodlands which cannot be included as agricultural land'.[41] The country's agricultural land was situated a fair distance away from the country's oil centres and oilfields. By 1948 the industry had acquired 9.2 per cent of the total land surface, but only operated on 0.14 per cent of this area. A further 6.2 per

Table 22   *Distribution of oil lands, 1948*

Land usage	Surface in hectares	% of total land surface	% of total oil lands
Surface area of general interest to the industry	8,498,700	9.32	100.00
Area in exploitation	131,528	0.14	1.55
Area not used	8,367,173	9.17	98.45
Potential oil-bearing land	5,665,365	6.21	66.66

Source: Walter Dupouy, 'El petróleo y las tierras agro-pecuarias', *El Farol*, 10:111 (Ago 1948), 2–9, Table 3, p. 7

Table 23   *Type of land on which concessions held, 1948*

Source of land acquisition	Surface in hectares	% of total land surface	% of total concessions
Land acquired by companies	163,007	0.178	1.92
Land leased from landowners	1,189,383	1.3	14.00
Land leased on communal land	173,289	0.19	2.03

Source: Walter Dupouy, 'El petróleo y las tierras agro-pecuaries', *El Farol*, 10:111 (Ago 1948), 2–9, Table 4, p. 7

cent of the total land surface was considered by the industry to be potential oil-bearing land, of which only 1.03 per cent was later acquired directly or leased from landowners, representing in all 1.5 per cent of the country's total surface area. During the 1920s these figures would have been smaller since not all the operating companies present in 1948 had entered the country.

However, it was the substantial wages earned by the oil workers that had people, especially landowners, complaining about the oil companies. In March 1924, for instance, it was reported that oil workers were receiving Bs.6 per day, 'which was detrimental for agriculture because the peasants were abandoning the countryside to work for the companies'.[42] As the companies intensified their work, so the shortage of manpower for agriculture became more acute. In 1927 many *hacendados* in Trujillo State were short of men, and in Carora 'prices of essential goods are going up because of a lack of labourers, of which at least 2,000 have migrated to Zulia'.[43] A similar situation was reported in Barlovento, Cumaná, Margarita, and around Maturín and Carúpano in eastern Venezuela where Exxon had important operations. In the same year many landowners in the interior were willing to sell their land because they could not hire labour to work their lands, nor could they afford to pay the same wages as the oil industry.[44]

Many modern writers, such as Brito, Rangel, and Quintero,[45] echoing the contemporary reports, have attributed the demise of the agricultural sector to *inter alia*, a rural exodus created by the high wages paid by the oil industry. It is true that wages in the oil industry during the latter part of Gómez's period in office were disproportionately high in comparison to other industries; for example, in 1927 the oil companies paid a total of Bs.77 million in wages and salaries to only 1.3 per cent of the population, compared to the Bs.130 million paid in agriculture, and the Bs.65 million paid in industry,[46] which meant that annual wages per worker in the oil industry were on average Bs.5,598, compared to Bs.3,692 in industry and Bs.209 in agriculture for the same year.[47] But these figures are meaningless unless we look at the overall wage structure of the industry and of Venezuela as a whole, during these early days of the oil industry. There is little doubt that the total wage bill for the oil industry was large; the question to determine is the effect that this had on the wages earned by unskilled labour in other parts of the economy.

It is first of all necessary to look at the composition of the total wage bill. In 1936 salaries (see Table 24) accounted for 39 per cent of the oil companies' total wage bill, of which 63 per cent was paid to foreign staff, who represented 48 per cent of the total number of employees. Accordingly, salaries for foreign staff at an average of Bs.965 per month were considerably higher than those paid to Venezuelan employees, which were on average Bs.482 per month. The picture is the same for manual workers: here foreign workers earned Bs.14.41 per day compared with Bs.9.0 per day for their native counterparts. While foreign workers comprised 5.6 per cent of the total number of workers in 1936, they received 9 per cent of the total wages. The oil companies provided their staff with subsidized accommodation and other facilities, so that it can be assumed that a large part of the income earned by foreigners was repatriated out of the country, thereby reducing the impact of the total wages paid to them on the economy. According to Brito, in 1912 wages in Zulia were on average Bs.1.25–1.50 for peons on a ten-hour day, with women and children receiving considerably less at Bs.0.75 and Bs.0.50 respectively. In central states wages were higher: in Aragua, for example, coffee pickers were paid Bs.2–2.50 a day. Despite the low wages and bad living conditions, the oil companies when they entered Zulia found 'great difficulties in recruiting sufficient unskilled labour for work'.[48] This scarcity affected other industries such as the sugar mills of Zulia which were forced to use Goajiro indians.[49]

The widespread use of indentured labour was another feature of agriculture during this period. One of the largest sugar plantations in Zulia, for example, recruited its labour in Caracas, offering good wages of Bs.150 a month in addition to free food and lodgings. Once the workers accepted they received an advance on wages of something between Bs.50 and Bs.100 which would later be discounted from their wage packet. When they arrived at the sugar mills they found that their real wages were Bs.2 instead of the promised Bs.5 per day, and because their initial debt had not been repaid they were forced to remain working at the mills. In addition, part of their wages were paid in kind, with the company store being the only place

Table 24 Distribution of wages and salaries in the oil industry, 1936 (in Bs.)

Classification	Wages and salaries	% of sub-total	% of total	Number of employees	% of sub-total	% of total
White collar workers:						
Foreign staff	14,681,766	63.0	24.0	1,268	46.0	9.0
Venezuelan staff	8,616,372	37.0	14.0	1,490	54.0	11.0
SUB-TOTAL	23,298,138	100.00	39.0	2,758	100.0	20.0
Blue collar workers:						
Foreign workers	3,164,436	9.0	5.0	610	5.5	4.0
Venezuelan workers	33,762,808	91.0	56.0	10,386	94.5	76.0
SUB-TOTAL	36,927,244	100.0	61.0	10,996	100.0	80.0
TOTAL	60,225,382		100.0	13,754		100.0

Source: adapted from Alirio Parra, La industria petrolera y sus obligaciones fiscales en Venezuela (Caracas, Primer Congreso Venezolano del Petróleo, 1962), Table 12, p. 73

where they could redeem the vouchers. In 1911 Gómez had given orders to his State Presidents to have this practice stamped out,[50] but it still persisted. Ifor Rees, the British Consul in Caracas, reported in 1915 that 'on some estates it is the custom to pay labourers in a kind of "estate money" which will only be accepted at the owner's store on the estate, thus obliging their labourers to buy their goods at these stores'.[51] Such systems persisted up to the time of Gómez's death.

Wages differed from region to region, and between sexes, but more importantly, the wage differential that allegedly attracted peasants to the oil camps was in existence long before the arrival of the oil companies in the form of wage differentials between the urban sector and the rural sector of the economy. Ifor Rees, for example, reported that in 1913 the average city wage was Bs.4.60 per day compared with a peon's wage of Bs.1.50 per day on a sugar or cocoa estate.[52] In addition, if the differences that existed between the mining sector of the economy and the agricultural sector are compared, it is clear that long before the arrival of the oil companies distinct wage differentials existed which exerted as much influence on the rural population as did the oil companies later on. The average wage paid in the mining industry in 1917 was Bs.8 per day, whereas that of CPC was Bs.6,[53] and by 1922 wages for the industry as a whole were compatible, at Bs.4 per day, with urban wage levels. By 1926, however, wages had increased to Bs.7 per day, which was above the average urban wage, and began to exert a slight pull in attracting rural labour. Certainly by 1936 there were very few industries in Zulia that could match the oil industry's average daily and monthly wages for native staff of Bs.9 and Bs.482 respectively. This is revealed by the 1936 industrial and commercial census of Zulia, from which it can be seen that the average monthly salary was Bs.344, and that the average daily wage was Bs.5.7, and that only three sectors, *viz.*, chemical firms, banks and insurance companies, and transort services, paid on average higher salaries than the oil companies.[54] Given the tremendous productivity differences between the oil industry, industry in Caracas, and agriculture (calculated by Rangel to be, in 1936, Bs.49,747, Bs.9,884, and Bs.616 per man employed respectively),[55] it is surprising that wages and salaries for the oil industry were not much higher.

Despite the oil industry's high wages, the number of migrants to the oil camps, estimated by Mieres at 60,000–80,000,[56] representing only 3–4 per cent of the rural population, was small. Using Brito's figures, it can be seen that migration directly attributable to the oil industry up to 1921 on average affected 5,285 people, of whom only a third (1,738) came from rural districts, with the rest arriving from urban centres.[57] Although migration increased during 1927–9 compared to the figures given above, the labour force in the oil industry remained on average, between 1925 and 1936, at 11,500 people per annum,[58] representing only 1.3 per cent[59] of the total active population in Venezuela in 1936 (compared to 67.4 per cent for the agricultural sector) and implying that the rural migrants who came to Zulia and other oil regions were employed in areas outside the oil industry, most probably in commercial establishments.[60] Zulia had traditionally attracted people: in 1908, for instance, State President José Ignacio Lares complained that there was a floating population of 5,000 people who were a considerable nuisance,[61] and in 1918 State President Santos Matute Gómez informed Gómez that the population was increasing on account of the 'migrants from other parts of the Republic and from abroad'.[62] The local effects of the oil companies were dramatic: for example, the Municipality of Lagunillas increased its population from 982 people in 1920 to 13,922 in 1936,[63] an annual increase of over 80 per cent. The great majority of these migrants came from Trujillo and Táchira, with the result that in 1927 these states began to restrict the flow of people to Zulia.[64] The oil industry also recruited a large number of black West Indians, but this source was curtailed when in December 1929 a government decree restricting black West Indian immigration was issued.[65]

Nevertheless, the migration to the oil states was small compared to what was happening in the rest of the country. According to Moore, most migrants (who counted for just under one-eighth of the population) during this period headed for Caracas and for the non-oil states and territories, with only 24 per cent of the internal migration shift accounted for by the petroleum states. Moore concludes that the direct effect of an increase in petroleum wages had little influence on wages elsewhere, but it did result in increased activity in other sectors of the economy such as services, commerce, manufacturing industry, and government employment.[66]

In 1936 the oil industry accounted for only 4.6 per cent of the country's national income, compared with 17.6 per cent for government, 20 per cent for manufacturing industry, and 21.5 per cent for commerce and services.[67] While the agricultural sector grew between 1925 and 1936 at an average rate of 3.5 per cent annually, industry and services expanded by an annual average of 11.3 per cent and 9.9 per cent respectively,[68] with the result that the majority of urban migrants headed for places which offered the possibility of employment: for example, according to Rangel, the government's bureaucracy between 1920 and 1936 rose from 13,500 to 53,100 staff, an increase of 316 per cent.[69] It can be concluded, then, that there was a much larger migration to the cities in search of general employment than to the oil or other related industries.

The drop in coffee prices from Bs.140 per quintal in 1929 to Bs.50 per quintal in 1933 produced a further decline in the country's agricultural export sector. The position was made worse by the refusal of the traditional commercial houses and banks to extend credit to the coffee producers, resulting in the loss of two-thirds of the 1930 crop because of a lack of funds for its collection and transportation.[70] The Banco Agrícola y Pecuario distributed a reasonable amount of money among the farmers of the country: for example, between July and December 1930 the bank lent Bs.6.1 million to 177 applicants, of which 44 per cent was used to pay off mortgages (Bs.2.8 million), 38 per cent to help agricultural farmers (Bs.2.3 million) and 18 per cent to assist cattle farmers (Bs.1.0 million).[71] Nevertheless, the plight of the coffee growers worsened as the economic depression deepened. The value of coffee exports fell from Bs.133 million in 1929 to Bs.61 million in 1931.[72] But other agricultural sectors also suffered from the worsening economic situation: for example, cattle exports to the Dutch Antilles, previously around 450 head of cattle per month, had fallen considerably by 1933[73] and low prices for other crops meant that many were not harvested because it was unprofitable to do so, with the result that many people left agriculture. The persistent low prices for coffee (by early 1934 prices had plummeted to Bs.26 per quintal) forced many people who had received loans from the Banco

Agrícola y Pecuario and other banking institutions to default.[74]

It was not only the coffee growers who faced ruin, as cattle ranchers and farmers alike of the western plains of Anzoátegui, for instance, appealed to Gómez for help.[75] The Coffee Growers' Commission, in a desperate move, approached Gómez to request the government to write off the interest charges on loans extended by the Banco Agrícola y Pecuario. Consequently, on 24 July 1934 the government consented to grant a free loan of Bs.10 million,[76] without any quid pro quo, to growers who could prove their need. The intention was to give the industry a boost by providing it with the wherewithal to bring in the harvest.[77] However, the growers, having secured loans without the need to provide collateral or repayment, considered these as net profit, and so did nothing towards their plantations. One immediate effect was for the sugar growers to request a subsidy of Bs.2.5 million for their industry,[78] and later a Cocoa Producers' Association was formed.[79] But the real problem was that the loans were not used to increase productivity.

The decline of the agricultural sector, although accelerated by the advent of the oil industry, was due principally to deficiencies created by low levels of investment, low productivity, and scarce technological resources. The political stability of the country, together with the growth of the oil industry, expanded local economies and placed a severe strain in places like Zulia on the supply of agricultural labour. But this rural shift to the cities, albeit on a small scale, had already started before the oil companies had arrived. Nevertheless, the oil companies' demands for labour, and the wages they paid (at first equal to the urban norm) exerted their effect on the local economy, but did not contribute significantly to the urban migration that took place over the whole country during the period in question. The suggestions that the industry acquired and retained a vast amount of cultivable land cannot be substantiated. The lack of adequate investment contributed in large measure to the decline of the country's agricultural sector. While countries such as Brazil and Colombia increased their investment in coffee, Venezuela's production remained 'under the same rudimentary form practised since the industry started in the country'.[80] In 1939 the average cost of producing 46 kilograms of washed coffee in Venezuela, without taking into account administrative costs and interest charges, was much higher than in competing coffee-producing countries. At a time of falling prices Venezuelan production was quite simply uncompetitive compared to its rival exporters. In one respect oil revenues helped to check the demise of the agricultural export sector during the Great Depression since it is doubtful whether the Banco Agrícola y Pecuario and the government could have loaned money to the coffee growers without the additional revenue entering the Exchequer from the oil industry. However, these funds were wasted as many of the *hacendados* who received loans used them to set up other more profitable businesses. These economic factors, together with the very bad working conditions, had the combined effect of releasing people from the land, with the result that the majority who left went to Caracas while around a quarter moved to the oil states.

THE BOLIVAR COMES UNDER PRESSURE

The direct effects of the oil industry were, with some regional exceptions, relatively small. But the direct impact which the industry had on the exchange rate of the country's currency was crucial for the agricultural export sector. Thus, between 1925 and 1935, the direct economic impact on the country's national income was an average 11.5 per cent per annum, the sum that this represented being mostly converted from U.S. dollars to bolívars in order to pay taxes, royalties to concessionaires, wages and other expenses incurred. This allowed the companies to manipulate the country's exchange rate to their own benefit, and to the detriment of the country's traditional exports.

As mentioned previously, during the early 1920s coffee prices remained very low, and had an adverse effect on the country's balance of payments. The resulting deficit was reflected in the bolívar's exchange rate, which was devalued in relation to the U.S. dollar by 6.5 per cent in 1920, 16 per cent in 1921, 5 per cent in 1922, and 1 per cent in 1923, only reaching parity again in 1924. The country was able to overcome the crisis because the oil industry had injected into the economy large sums of money. The bolívar was, therefore, artificially maintained at parity, which only stimulated imports and did little to alleviate the country's problems. As long as oil production expanded, the government had a greater margin for error in balancing its accounts. According to Knudson, government expenditure in 1917–18 exceeded the projected budget by 25 per cent, whereas in 1924–35 this imbalance had increased to 100 per cent, with the result that the 'original budget was practically meaningless'.[81] Consequently, the oil companies' curtailment of production in 1927 had an immediate effect on government revenue, and on real demand in Zulia and, to a lesser extent, in the whole country. Thus, government revenue from the oil industry and from customs receipts fell resulting in 1928 in a budget deficit which, by 1930–31, had increased to Bs.51 million.[82] Many of the credits which Venezuela had been promised did not materialize, and in order to balance the books the country's gold reserves were dropped by 50 per cent. The measures introduced by Gómez to cut down government expenditure only produced further deflationary pressure on the economy. The *Boletín de la Cámara de Comercio de Caracas* interpreted the crisis partly as a result of ineffective measures taken to curb the country's problems before the oil boom, when the country preferred to continue its 'carefree existence of depressingly expensive pleasures, and calling it progress in order to appease any guilty feelings'.[83] The situation was further aggravated by Venezuela's lack of a central bank, which could have controlled the supply of money and the value of the bolívar. Six leading private banks were authorized to issue bank notes (which had to be backed 100 per cent with gold), and the Banco de Venezuela acted as the government's fiscal agent. According to Knudson, the bolívar was probably the strongest gold-backed currency in the world at the time, but a strong currency did not ensure that an adequate money supply was maintained because unsupervised commercial and financial transactions drained the economy of its liquidity, exerting a deflationary effect on the economy. The lack of a central bank left the bolívar's international exchange rate to be determined by the commercial transactions of the private banks. Because of a deteriorating balance of payments the bolívar depreciated in relation to the U.S. dollar, and the government was unable to maintain parity since it could not intervene in the market. In addition, the large number of dollars which had previously maintained a strong bolívar had decreased due to a decline in oil production. This benefited the oil companies which had a vested interest in maintaining a high rate of exchange with the dollar 'as this represented a reduction in what they had to pay the government'.[84] The higher the exchange rate the fewer dollars the companies needed to spend to pay their expenses in Venezuela. For example, if the companies' tax bill amounted to Bs.1,000 per month, the companies would need to sell $192 if the exchange rate were Bs.5.20 per U.S. dollar, but if the exchange rate went up to Bs.6.10 per U.S. dollar, then the companies would only need $164 to meet their tax commitments, an effective tax reduction of 14.6 per cent.

As the economic depression deepened so did the bolívar depreciate further, reaching an exchange rate of Bs.7 to the U.S. dollar in 1931. In part this was due to speculation, and to a low import tariff on agricultural goods which allowed cheap food imports into the country. At the same time there was a growing feeling among certain sections, especially the native banking community, that the foreign banks, through a massive repatriation of funds, had depreciated the bolívar considerably.[85] A growing number of people, headed by Bernardo Jurado Blanco, called for the nationalization of all foreign banks.[86] Although it is true that the transactions effected by the oil companies benefited the foreign banks, it is doubtful that foreign banks were to blame directly for the bolívar's depreciation. The currency's devaluation was due, as the *Boletín de la Cámara de Comercio de Caracas* pointed out in 1931, to an unfavourable balance of payments which made the exchange rate fluctuate, and 'nobody can prevent people holding dollars to benefit from the situation'.[87]

As the crisis worsened the government remained conspicuously inactive, secure in the belief that market forces would establish the right exchange rate. But the international nature of the oil companies and of the foreign banks operating in the country meant that the companies' dollar transactions could take place in New York or London, with the local branches of the banks in turn supplying the operating oil companies with their cash needs. The use of this method curtailed further the supply of dollars entering the Venezuelan economy, and thus aggravated the alteration in the exchange rate. In order to maintain the high exchange rate, the companies sold part of their foreign drafts in the country. At times, when the Venezuelan banks such as the Banco de Venezuela would refuse to take the drafts because the price was too high, the companies would then offer them directly to the trading houses which were desperate for foreign currency, thereby forcing the banks to buy foreign drafts. This occurred in March 1932 when seven banks decided to buy dollars at Bs.6.40, and sell them at Bs.6.45. As Sanabría argues, both the oil companies and the foreign banks had a common interest in maintaining a depreciated bolívar, and thus neither felt obliged to take into account the country's interest in the matter.[88]

Despite Gómez's belief in market forces, there were close advisers who wanted the government to take positive action to deal with the crisis. Abel Santos, for instance, proposed that Congress reconvene at an Extraordinary Session to debate the crisis and to take appropriate action. But Gómez considered that such action would only provoke further alarm.[89] From Washington, Arcaya mounted a vigorous campaign to move the government to demand payment of oil taxes in gold specie,[90] and although he agreed with Gómez's *laissez-faire* policy, he thought that the oil companies should not benefit from the country's depreciated currency. He calculated that the companies in 1931 had saved $2 million in taxes because of the devalued bolívar. The Venezuelan government, led by Pedro Rafael Tinoco, the Minister of the Interior, had serious doubts about the desirability of such a form of payment since the companies for their part could equally demand gold specie in exchange for the bank notes they held, with the possibility that such action might precipitate a run on the country's currency. To Arcaya such fears appeared fantastic 'because it is not obligatory to redeem small denomination notes in gold';[91] moreover, the government could avert any controversy by simply extending loans to foreign banks in gold, part of which would have been received from the oil companies.[92] The situation worsened when the U.S. left the gold standard in 1933, and when Venezuela followed suit in July 1933. According to Arcaya, the government, through the depreciation of its paper money, and through its failure to insist that the oil companies pay in gold specie, had, up to October 1933, lost $4.5 million, in addition to the losses it had incurred by having to acquire foreign exchange at a premium in order to meet its foreign payments.[93]

The bolívar's devaluation favoured the oil companies, but it also alleviated slightly the plight of the agricultural exporters, who found that their products were now relatively cheaper to sell in foreign markets. It was the large trading houses which bore the brunt of the devaluation because the prices of imported goods had increased accordingly. The various problems faced by the three sectors of the economy (oil, agriculture, and commerce) were discussed by various representatives in June 1932, but no agreement was reached.[94]

By 1933 the bolívar had strengthened on account of a reduction in demand for foreign exchange to service foreign debts, a depreciation of foreign currencies, especially the U.S. dollar, and an increase in oil activity. This appreciation of the currency meant that the comparatively small advantage which agricultural exports had gained on the world market during the previous years soon disappeared, creating internal problems for Gómez. The increase in oil production, as well as the devaluation of the U.S. dollar when it was taken off the gold standard, resulted in the revaluation of the bolívar, which in 1934 reached a fluctuating exchange rate with the dollar of Bs.3.00–3.05. Coffee and cocoa exports suffered heavily, while at the same time imports of other agricultural products increased since they were relatively cheaper than those produced in the country. The question of devaluing the bolívar was mooted, but while this would benefit the agricultural exporters, it would be of the greatest advantage to the oil companies. The oil companies' pay schedule in 1933 amounted to $34 million, and by devaluing the bolívar by half the companies would in effect have to pay only $17 million, which would result in a real loss of revenue to the country.[95] It was therefore decided to introduce a two-tier exchange rate, known as the Convenio Tinoco, along the following lines: the oil companies agreed to sell their dollars at Bs.3.90, and any surplus to the government at Bs.3.03. The banks in turn agreed to sell dollars at the rate of Bs.3.93 and sterling at the rate of Bs.20 in place of the free market rates of Bs.3.07 and Bs.15.40 respectively. The government benefited from this transaction, too, because it would acquire dollars at

Bs.3.03 and sell them for Bs.3.93.[96] The government's intention was to purchase gold with the profits from such transactions, and to transform all of its reserves into gold reserves.[97]

The system was designed to help the agricultural sector, but the oil companies achieved a devaluation of almost a third in the value of the bolívar, provided that there was a willingness to pay the higher exchange rate. It was soon apparent, however, that importers were unwilling to purchase dollars at the devalued rate, and the government was obliged to take all the dollars (about Bs.3 million per month) offered by the oil companies. The companies, therefore, did not achieve the gains they expected from the agreement, which placed the Treasury in an awkward position as it would be unable to purchase dollars indefinitely. Arcaya counselled from Washington that the stability which it was felt could be achieved through the agreement of the previous August between the banks and the oil companies was 'unlikely'.[98] In New York the exchange rate was provisionally set at Bs.3.35 instead of the Bs.3.90 fixed in Caracas. This difference gave rise to speculation and fostered a lack of confidence. According to Mieres, the exchange rate of Bs.3.35 was finally adopted,[99] allowing the oil companies to make a modest saving of around 10 per cent on their tax payments, while at the same time allowing imports, especially of foodstuffs, to increase. In 1937 a new agreement was signed with the oil companies in which they agreed to sell dollars at Bs.3.09.

The oil companies took full advantage of the adverse exchange rate which resulted from the country's balance of payments difficulties and which they were favourably placed to manipulate, within certain limits (as was seen during the early 1930s), to suit their needs. This was to be expected, for oil companies are in business to achieve the best possible return on their capital. But many leading figures in the country and close advisers to Gómez also expected that the government would take prompt action to solve the problems. However, the delay in reaching a solution meant, as in the case of agriculture, that it would come too late to be of any real value. It is true that, during the period leading up to the imposition of an import tariff on oil by the U.S., Gómez was unwilling to upset the oil companies, lest they should abandon or reduce their operations to a minimum, but this should not have prevented the government from acting more vigorously. The lack of a central bank severely hampered the government in managing the economy, but it is doubtful that Gómez foresaw the disruptive impact the oil industry would have on the country, and therefore did not fully grasp the urgency of arriving at a quick solution. Nevertheless, Gómez's innate liberal economic outlook prevented him initially from interfering with the country's economy.

The economic structure of the country was unprepared for the economic boost by the oil industry, and the increases in effective demand, instead of leading to capital investment, to a deepening of the nascent industrial activities, and to a strengthening of agricultural production, were channelled into higher levels of consumption of imported goods. Although the oil companies, through their domestic wage structures, could influence to some extent the general rhythm of economic expansion and stagnation, they nevertheless did not generate nor assist in the formation of a strong industrial base since the value which they added to the oil in Venezuela was minimal as they preferred, for political reasons, to establish their downstream operations outside the country. Consequently, once the industry was established, an increase in crude oil demand had little effect on the country's economy, apart from the increased oil revenues to the government.

The increase in commercial activity in the country meant that manufacturing industry suffered a decline during this period, for people preferred to invest in commerce which gave a rapid profit. However, the high level of imports, made possible by the large foreign reserves accumulated from the oil industry, created a severe strain on the balance of payments when the companies reduced their level of operations in the late 1920s. But the increase in income levels in the country should have had a healthy effect on agricultural production. However, the industry was unable to respond vigorously enough, and increased demand was satisfied instead by imports of foodstuffs, which dissipated any effect the former might have had on domestic output. Although it is true that the companies had an effect regionally on agriculture, severely disrupting agricultural output and creating a scarcity of labour, this only served to accelerate the decline which had started before the advent of the oil industry. Nevertheless, the low exchange rate for the bolívar during the mid-1930s on account of a temporary lull in oil production, did make agricultural exports even more uncompetitive in world markets.

The increase in government revenues from oil taxes allowed the government to accelerate repayment of its foreign and internal debts, as well as to integrate the country's diverse regions through the construction of a primitive road network. At the same time the government gained greater freedom to act as it lessened its dependence on customs receipts which were a direct function of the buoyancy of the economy. But by the same token the government did not deal adequately with the important problem of the decline in agricultural production, all the more surprising given Gómez's 'telluric' background. The problems were recognized, and suggestions made to remedy them, long before they reached the level of a crisis, but Gómez failed to act decisively, believing that market forces would determine the best possible solution, a belief which proved, in the end, both wrong and costly.

LOCAL EFFECTS OF THE OIL INDUSTRY

The wealth generated by the oil companies had a relatively small impact on the general economic conditions of the country, but its socio-economic effects on Zulia in general, and on Maracaibo in particular, were startling. In Maracaibo in late 1925, for example, two oil companies alone accounted for weekly purchases of Bs.500,000.[100] In the same year there were on average between 25 and 30 ships of 1,000–3,000 tons per month plying the route between Maracaibo and Curacao.[101] New towns flourished around the oil camps at Mene Grande, La Rosa, El Mene, Rio de Oro, Tarra, La Concepción, La Paz, and San Lorenzo, with populations varying between 1,000 and 10,000 people.

Maracaibo, as the capital of Zulia State, had always been at the forefront of progress in the country, but it was ill-suited and ill-prepared for the oil industry. Its harbour, which was only 100 metres long, was in sad disrepair, and could not cope with the large volume of imports. Outside the camps, the general conditions were bad: the city had practically no sewage, there was no adequate aqueduct to supply it with fresh water, and the state of the roads was poor. Property values and rents in Maracaibo increased to astronomical levels. Houses which in 1914–18 had cost Bs.20,000 were being sold for Bs.200,000 in early 1926.[102] Land which previously had fetched Bs.400 per hectare in Bella Vista, where most of the oil companies had established themselves, sold in 1926 for Bs.15,000–25,000 per hectare.[103] Nevertheless, some 100 houses clustered into camps were being constructed for the oil companies there.[104]

LOCAL GOVERNMENT

The trading houses in Maracaibo were the main beneficiaries of the new-found prosperity. The amount of speculation that took place induced P. León, the Secretary-General of the Zulia government, to advise Gómez of the need to 'suppress with a strong hand those government officials who buy and sell everything, including their honour, liberty and justice'.[105] This state of affairs had begun long before the advent of oil; for example, Alberto Aranguren's first measure as the Interim State President in April 1916 had been to 'abolish the contracts and monopolies'[106] his predecessor, General José María García, had allowed to flourish. Nevertheless, the monopoly concessions and illicit business deals continued. In 1919, for instance, José J. Gabaldón, President of Zulia, reported that the meat monopoly held by Christern Zingg & Cia in Maracaibo had had a detrimental effect on the supply of good fresh meat to the city. Gabaldón rescinded this monopoly contract, and the meat trade was normalized in 1919 'on the basis of complete freedom for the industry'.[107] But as Gabaldón pointed out, this was only possible because he could supervise the trade himself. In the other eight districts which made up Zulia State, the *Jefes Civiles* and the *Comisarios de Caserío* were a law unto themselves. In 1920 the *Jefe Civil* of the Municipality of Bolívar kidnapped the wife of a foreigner, 'plunging a distinguished home into shame'.[108] In 1922 the Palmarito Sugar Co., an American concern, charged that Colonel Emilio Abreu, *Jefe Civil* of the Municipality of Independencia of Libertador District of Mérida State, had built a house on company property, and that he had requested the State's President to grant him 1,000 hectares which belonged to the company.[109] Albino de J. Medina also complained that the *Jefes Civiles* of the municipalities of the Colón District abused their powers by imposing arbitrary fines.[110] Gómez therefore appointed M. Toro Chimies, who was the Secretary-General to Gómez's cousin Santos Matute Gómez, the President of Zulia, to correct these and other complaints. Gómez wanted Toro Chimies to 'act wisely'[111] in order to avoid any complications and 'international disputes, which could arise from the treatment of the large number of foreigners in Zulia as a result of the mining activities taking place'.[112]

LABOUR UNREST

Gómez was well aware that labour relations in the oilfields were not good. The companies had acted with little concern for local sensibilities; for example, much oil from the producing wells and from wells which were being brought-in spilled over Lake Maracaibo causing many fires. Great terror was instilled in the local population but the companies refused to recognize any claims for damages arising from the fires. In 1925, however, opposition to the oil companies came directly from the oil workers who went on strike demanding higher wages and better working conditions. Although the Venezuelan Gulf Oil Co. had instructed all its drilling operators in June 1925 that local employees earning more than Bs.10 per day would be obliged to sign a form 'renouncing any compensation'[113] for accidents or unfair dismissal, it was VOC which was singled out for its contemptuous treatment of its workers.[114] Thus, on 4 July 1925, a strike, headed by Antonio Malavet, was declared at the Mene Grande camp of VOC, which later spread to other camps of the same company, and to Lago and the Venezuelan Gulf Oil Co., affecting a total of 2,500 workers. The strikers demanded a minimum wage increase from Bs.6 per day to Bs.8 per day, with a standard 8-hour day.[115] On 21 July the workers at La Rosa went on strike demanding a minimum wage of Bs.10 per day.[116] The companies appealed to General Febres Cordero, the President of Zulia State, who said that he could only send 30 armed men to protect the oilfields.[117] The companies then appealed directly to Gómez for help, but he was occupied in

Caracas with a number of industrial disputes which were taking place at the time, and which appeared to have been government inspired.[118] When Gómez did not reply to the companies' appeal, they were forced to re-appeal to Febres Cordero, and to reach an agreement with the striking workers. Minimum wages were to be increased to Bs.7 per day,[119] and when Febres Cordero in turn promised 'full protection' to those workers who wished to return at that rate of wages,[120] the trouble soon ended.

SOCIAL PROBLEMS

The social problems between the oil companies and the population remained. There was growing resentment at the disorderly and drunken behaviour of a large section of the foreign oil-workers. But as Wainwright Abbott, U.S. Chargé d'Affaires, pointed out in 1926, the local population had been quick to seize the opportunity to overcharge foreigners, with the result that quarrels had resulted.[121] Although many of these were petty, in numerous cases the welfare of an entire community was at stake because of the companies' abusive behaviour. The case of the town of Cabimas illustrates the local effect the oil industry had on a small community in Zulia.

In 1920 Cabimas had a population of 2,000 which by 1936, had increased to 22,000 (a 1,000 per cent increase).[122] The oil boom changed Cabimas from 'its peaceful bucolic atmosphere, to a disordered amalgam of human beings, giving it the appearance of a gypsy's shop'.[123] The lack of any effective law and order led the drillers to clash continuously with the local authorities. For example, on New Year's day, 1926, one of Lago's employees and a number of other foreigners from VOC were returning from La Rosa to Ambrosio after a night's celebrations. Just before entering Cabimas they were detained by a group of men who, according to their lawyer, were pretending to be policemen.[124] As the oilmen were drunk, it seems likely that they were causing a breach of the peace, and the local group, composed of the Secretario de Juzgado of Cabimas, a lawyer, and two civilians, tried to detain them in Cabimas because of their unruly behaviour. The oilmen, however, refused to recognize the group's authority to detain them, and so fisticuffs took place, with the Secretary and the two civilians being hurt. The police arrived and arrested one of the oilmen. The following day the *Jefe Civil* refused to release the oilman until his two companions came forward. When they did, they were promptly arrested.[125] All were released after fines of Bs.2,000 were paid. A few days later a truck driver from Lago was detained because of speeding. The company's foreman tried to get his release, but was unable to do so because he too was detained for speeding, and later arrested. Other employees of the company were arrested for minor infringements of the law and only released after payment of large fines.[126] The oil companies continued to have little regard for the needs of the local community, and for the effects that their drilling activities had on the local population. In 1925 L. Calvani, Inspector Técnico de Minas, reported that, since Lake Maracaibo was contaminated with oil, the companies should supply the lakeshore communities with fresh water.[127] The oil spillage over the lake meant that communities such as Cabimas, which depended for their supply of fresh water on the lake, now got water that was brackish from pollution with oil. Complaints by the population about the water supply, and about the proximity of VOC's drilling rigs to the town, were lodged with Gómez. After receiving Febres Cordero's eye-witness report,[128] Gómez requested Antonio Alamo, the Minister of Development, to order J.M. Braschi, the Inspector Fiscal de Hidrocarburos for Zulia State, to further investigate these complaints.[129] Braschi confirmed that the lake water around Cabimas was polluted, but that the drilling rigs were over 500 metres away from the town. His recommendations were that the companies supply the town with fresh water, and that no new wells be drilled less than 500 metres from it.[130] Nevertheless, the problem persisted as the company felt that it had the legal right to drill anywhere in Cabimas since its concession covered the town. Consequently, it placed a drilling rig next door to Clementina Romero's house, who complained to Gómez that the rig had forced her and her neighbours to abandon their property because it represented a fire hazard. The company had also forbidden the use of domestic fires in order to minimize the risk of fires. When the company refused to acquire Clementina Romero's house for Bs.20,000 or pay damages, she instituted a 'defence interdict' which was granted. The company, however, ignored this and

continued drilling, and later found oil which flooded her house
and the vicinity with five inches of oil.[131] It then successfully
obtained an eviction order against Clementina Romero, but she
complained that her point of view had not been adequately
heard because of the influence the company exercised in the
region, and appealed directly to Gómez for help.[132] Gómez
enjoined General Febres Cordero to assist Clementina Romero
in her legitimate claim for damages, but he appears to have
ignored the request.[133] The company also refused to pay
damages to people maimed in industrial accidents or to
dependants of those killed in similar circumstances.[134]

APPOINTMENT OF PEREZ SOTO AS STATE PRESIDENT

The problems which faced the State of Zulia were not all directly
related to the oil industry. The social conditions and services
offered by the state were appalling, and, according to León, the
amount of wheeling and dealing indulged in by the local
authorities was enormous. Moreover, there was a great deal of
distrust between the various sectors of society and the local
authorities. In the oilfields it was also necessary to have con-
scientious government officials able to maintain law and order
with 'a spirit of justice which punishes and does not speculate,
and who can limit themselves to their duty, thus forcing others
to comply with theirs'.[135] As a first step Gómez decreed on 10
April 1926 the refurbishment of Maracaibo's harbour, and the
construction of a new aqueduct.[136] It was also apparent that
Febres Cordero was not the person to bring order to the state.
For example, in February 1926 he incorrectly informed Gómez
that, apart from the gambling *remate*, there were no other
monopolies in the state.[137] Moreover, he was inept at governing:
for instance, the contract awarded by him for the collection of
refuse was an utter failure after only one month in operation.[138]
When further disturbing rumours reached Gómez that the oil
companies were financing a secessionist movement in Zulia,[139]
he decided that a stronger man was needed to bring the state
out of the chaos, and appointed General Vicencio Pérez Soto as
head of the state. Pérez Soto was entrusted with the difficult
task of re-establishing the rule of law, of dealing with the
various monopolies that constricted trade, and of restoring to

the municipalities the communal land lost to the oil companies.[140] Gómez expected Pérez Soto to accomplish this assignment within six months, and as an inducement gave him Bs.100,000 in May 1926.[141] On 7 June 1926 Pérez Soto took up his appointment, and immediately set to work to gain regional approval[142] in order to ensure that foreigners would never find any source of unrest to exploit for their own ends.[143]

PEREZ SOTO TACKLES THE OIL INDUSTRY

Within three months Pérez Soto could claim to have ended the pernicious monopolies that had constricted trade, started organizing a more honest local government, cleaned up the city, both morally and materially, started refurbishing the city's public buildings, reduced black immigration by 75 per cent, and expelled twenty undesirable foreign ruffians.[144] The oil industry, however, was a specially difficult task for Pérez Soto, who, on his arrival in Zulia, found the companies, especially the American ones, 'rather insolent, that is without due respect to the regional government, refusing to pay state and municipal taxes, seeking to ignore everything by fixing it with open cheques, flattering our greed, but depressing our moral entity'.[145] Oilmen were disrespectful to local authorities, especially during that year's 4 July American Independence celebrations.[146] They had illegally dispossessed municipalities of their communal land, had driven their cars and lorries without number plates and refused to pay taxes, and had also refused to pay import duties on foodstuffs. When Pérez Soto brought these problems to the attention of the companies' management, he became alarmed by the power which the managers felt they had in the country. Nevertheless, he succeeded in making a compact with Gulf Oil in which the company agreed to pay duty on foodstuffs imported, to ensure that cars and lorries carried proper documentation and paid road taxes,[147] and he persuaded Lago not to acquire Toas Island in Lake Maracaibo.[148] At the same time he appointed a commission composed of Jesús María Nava, Henrique Acosta, and a land surveyor named Fuenmayor, to draft a bill on the territorial division between the state's districts. The inter-district dispute had gone on for a long time, and was particularly irritating to oil companies which held concessions covering part or the whole of a district. The companies were therefore very pleased with the establishment of the commission, and agreed to 'comply with the dispositions set by the new law'.[149] Another problem which Pérez Soto managed to resolve amicably was that between the British Controlled Oilfields Ltd and several landowners in Zulia and Falcón. In 1922 Santos Matute Gómez contracted with the company for the repair of the Zulia–Falcón road which its tractors had partially destroyed in ferrying equipment from Zulia to its operation sites in Falcón.[150] Later the company built a road between Puerto Altagracia in Zulia and Mene de Mauroa in Falcón, and the State Government authorized it to acquire many of the farms through which the road was to pass. The small landowners deemed this unfair and claimed damages. Although the farmers had complained to Pérez Soto's predecessor, Febres Cordero, he had been unable to reach an amicable solution. Pérez Soto, however, managed to get the company to distribute Bs.70,000 among the landowners as compensation.[151]

But solutions to other problems were not so easy. On his arrival in Zulia, Pérez Soto had received many complaints from workers who alleged unfair dismissal from the oil companies. He was unable to rectify this situation because he did not have the legal powers to compel the companies to pay compensation.[152] The situation was further complicated by the fact that in most cases the companies hired their itinerant labour from intermediaries. The companies in these cases argued that the intermediaries were responsible for the welfare of the workers.[153] The only alternative for the worker was to pursue his claim for damages through the courts, a long, tedious and expensive procedure for a humble worker. A solution to the problem, Pérez Soto suggested, was that he, as State President, be vested with special powers to intervene in these cases so as to render a quick solution.[154] Instead, on 15 July 1926, the government appointed Rómulo Faría Nones to inspect the living conditions of the workers in the Zulia and Falcón oilfields.[155] Pérez Soto also drew Arcaya's attention to the inconsistencies of the companies' treatment of local and foreign staff. He wrote stressing the very good accommodation provided for foreign staff compared to that offered to local employees, which in some cases only amounted to a hammock slung between two trees.[156] Even worse, the companies, especially Lago and VOC, were replacing Venezuelan workers (because, according to the companies, of their inefficiency) with *chinos*.[157]

The major dissatisfaction with the companies, however, was still their treatment of the local community. In June 1926 R.A. Mora of Cabimas complained to Gómez that the oil companies still refused to supply his community with fresh water,[158] and in August of the same year the new *Jefe Civil* of Bolívar District, Pedro Pinto S., confirmed to Pérez Soto that Cabimas, La Rosa, La Salina, San Ambrosio, and Pueblo Aparte lacked water for domestic use because of oil pollution of the lake. The oil companies had, in compliance with article 68 of the 1925 Oil Law, established a number of water taps, but these were grossly insufficient to cope with the needs of the towns.[159] The Minister of Development, Alamo, was consulted, and was of the opinion that the law could be more rigorously applied to force the companies to provide a constant source of fresh water.[160] Following this, Lago informed Leonte Olivo, Pérez Soto's Secretary-General, that the company was installing a further water tap at La Rosa, Cabimas and Ambrosio,[161] and VOC announced that it would provide Cabimas with fresh water from its own storage tanks.[162] There were other minor irritants: for instance, in 1928 the tradesmen of Lagunillas complained that the Venezuelan Gulf Oil Co. had blocked one of the streams which was used to transport people and merchandise, and that the obstruction, if not removed, would cause severe flooding during the rainy season.[163] The company, however, accepted liability, and proposed to deepen the stream bed.[164]

One of Pérez Soto's greatest tasks was to solve the increasingly difficult question of the communal lands of the state. According to Pérez Soto, during the preceding administrations of Santos Matute Gómez and Febres Cordero much of the municipalities' communal land had been sold to other parties, and if any Junta Comunal opposed such a sale they were threatened with imprisonment.[165] Consequently, on his arrival in Zulia, Pérez Soto stopped all registration of land which was formerly communal, and in the case of Bolívar District, renegotiated a contract for the land which the Junta Comunal had leased to VOC.

Making use of the powers vested in its contract, VOC on 25 June 1924 concluded a two-year agreement with the Concejo Municipal of Bolívar District to lease the remaining communal lands at a rent of Bs.1 per hectare.[166] In the following year the composition of the Concejo Municipal changed and Régulo Reyes was elected its President. On 15 June 1926 he wrongly informed the council that the President of Zulia desired that the new contract proposed by VOC be approved by the council. The council voted against the contract and resigned in protest against this outside interference. A new Concejo Municipal was appointed by Pérez Soto,[167] which on 25 June of the same year ordered that no further construction of houses would be allowed on common land.[168] Acording to Arcaya, the Minister of the Interior, and Villegas Pulido, the Attorney-General, the sale of common land was illegal, with the result that Cabimas was able to reacquire the land it had previously sold.[169] The situation with regard to the common lands of Lagunillas was more difficult. The Municipality of Lagunillas had acquired its common lands in 1896 and worked them until 1922 when, as the result of a putative legal transaction[170] the lands were expropriated, becoming part of the property of La Tasajeras, owned jointly by Bladimiro Jugo Padrón, Betulio Guijarro, and General Santos Matute Gómez.[171] In this way, the manufacture of grass mats from the rushes of the Lagunillas swamp, the village's main occupation, was monopolized by the new owners, and although they allowed the inhabitants to continue producing their grass mats, the villagers were now compelled to sell only to one person and at very low prices.[172] Later La Tasajeras was divided in two, one half going to General Santos Matute Gómez (which he sold to VOC for £500,000), and the other half to Betulio Guijarro and Bladimiro Jugo Padrón (on which the company took out an option to buy for £250,000).[173] When it became known that Pérez Soto intended to return the communal lands to the village, General Santos Matute Gómez, through his attorney Eduardo Ramírez López Méndez, offered to sell his share back to the Concejo Municipal, and Guijarro and Jugo Padrón offered to compensate the people of Lagunillas by providing them with alternative land on which to build a new town.[174] Pérez Soto was satisfied that this would save legal costs, and be equitable to all concerned, since Lago and the

Venezuelan Gulf Oil Co. intended drilling near the village and Pérez Soto had already proposed moving it to a new site. At the same time he helped the Concejo Municipal to draft a reply to the offer of a new contract by VOC. Carlos Dupuy, the new President of the Council replied that the Aranguren contract which the company held did not confer the right to exploit the communal lands, and that therefore the 'municipality repossessed the common land rented to the company, and does not recognise the company's right to demand that the council "refrain" from processing requests for common land'.[175] The council was, however, willing to consider a new leasing agreement with the company which, after further negotiations, was reached on 13 October 1926 when the company consented to increase its annual rent from Bs.1 to Bs.2 per hectare. In addition, the agreement stipulated that the company reserved the right to determine where new houses were built, and retained the oil rights over the land. This meant that the company could compel a houseowner to move, after it had paid suitable compensation, if it wanted to drill over the area.[176] Pérez Soto was pleased with this arrangement, and informed Gómez that due to the poor quality of the land the 'municipalities will never secure a better tenant than the company'.[177] Later, VOC agreed to compensate the people of Lagunillas with 5,000 hectares, personally selected by Pérez Soto, to form its new communal lands.[178] But Pérez Soto's problems with the oil companies had not ended as the question of the relocation of Lagunillas still remained.

In August 1926 both Lago and the Venezuelan Gulf Oil Co.[179] requested from Pérez Soto the relocation of that part of Lagunillas situated over Lake Maracaibo in order to prevent a major catastrophe during new drilling operations in the area. The companies reasoned that the oil spillage from their wells and the fires used for cooking in the village constituted a serious fire hazard both for the two companies and for the people of Lagunillas.[180] While he consulted the government, Pérez Soto requested the companies to suspend their drilling work so as not to endanger life and property. Nevertheless, he warned Gómez that a delay in reaching a solution could mean trouble as the companies were getting impatient.[181] The companies' legal step would have been to institute expropriation proceedings against the village, but such legal action would undoubtedly have had adverse political effects on the government since it would appear that two foreign companies were expropriating a Venezuelan village; furthermore, legal proceedings would take a long time, thus hampering the development of the oil industry in the region. Pérez Soto proposed that he should act as an intermediary between the companies and the people of Lagunillas in order to reach a quick solution,[182] and that the village should be moved to a healthier site. The same communal land as before would be granted, and the companies would pay the costs of the move. The advantages of such a scheme were that the town would be better planned, and that it would be given better resources with which to continue the manufacture and trade of grass mats. This would be the most equitable and conciliatory solution possible since, as Pérez Soto explained to Arcaya, there was a potentially large oilfield under the village. The cost of moving the town would be a third of the value of the prospective oil found in the area. Consequently, he ordered Pedro Pinto, *Jefe Civil* of Bolívar District, to halt all further construction in Lagunillas in order not to extend further the list of people claiming compensation.[183] But Gómez and Arcaya disapproved the plans, and advised Pérez Soto on 1 October 1926 that he should not intervene in the dispute, that the companies' request to have the village moved had been turned down, and that all work near Lagunillas was to be indefinitely suspended. If the companies still persisted in their wish to drill near Lagunillas, they had to 'settle their rights before the Development Minster and the Courts'.[184] This was a ruse used by Gómez to obtain a better deal for the people of Lagunillas, for on 11 October he informed Pérez Soto in a coded telegram that he should encourage the residents of Lagunillas to seek high prices for their dwellings, and to build new ones, 'in order to place obstacles in the companies' path'.[185] The companies appealed to Arcaya for a solution to their problem,[186] for they maintained that the residents of Lagunillas had no legal right to build their homes over the surface of the lake. However, spurred on by Gómez's encouragement, the residents of Lagunillas continued to build more houses over the surface of the lake. The government reached a decision in early 1927, as a result of which the village would remain situated in the same place, but no further construction would take place.[187] In order to allow the companies to buy up more houses over the lake, Arcaya purposely delayed dictating the Resolution regulating the government's decision. In June 1927 Arcaya finally authorized Pérez Soto to enact the necessary local regulation,[188] which was done on 15 July of the same year, detailing the new limits to Lagunillas within which construction could take place.[189]

After ten months in Maracaibo, Pérez Soto was able to announce triumphantly that 'Zulia is yours, General',[190] assuring Gómez that the secessionist tendencies had had their origin among foreigners spurred on by the great riches of the region, and by its ineffective government.[191] He was quite sure, after observing all the classes in Zulia very carefully, that it would not become 'Venezuela's Cataluña',[192] as many had predicted.

POLITICAL PROBLEMS

The increased mobility of people in Zulia and between the Dutch West Indies and the mainland was a constant source of worry to the local authorities, for it made the control of political activities and of the dissemination of political ideas and literature against the regime more difficult. Pérez Soto's political surveillance of the state was, with the active help of the oil companies, reasonably good. An agreement had been reached between the companies and the state government to collaborate in the surveillance of its foreign and native workforce, and to keep a close watch on any subversive propaganda that took place in the oilfields and in the camps.[193]

There were also renewed efforts to associate Pérez Soto with a fictitious Zulian secessionist movement. P.M. Reyes, a Colombian from Paris, notified Gómez that a Venezuelan lawyer, Gustavo Manrique Pacanins, had stated to him that important oilmen with close connections with the State Department had enquired about the suitability of Pérez Soto as President of Zulia.[194] Further reports on the creation of a Republic of Zulia by U.S. oil interests appeared in exiled opposition newspapers,[195] and Acisclo Boscan, on exile in the U.S., accused Pérez Soto of secretly plotting with Alejandro Rivas Vásquez (an enemy of Gómez living in Cuba) to create such a republic.[196]

But Pérez Soto dismissed these rumours as 'treason against the Fatherland, and an immense dishonour'.[197]

The increased prosperity of the Dutch West Indies also meant that many Venezuelan workers were employed on the islands, mainly by the large refineries. The excellent cover afforded by the large number of Venezuelans, and the proximity of the islands and their easy access to the mainland, had historically attracted to the islands exiled political groups who viewed them as a natural starting point for revolutionary activity in Venezuela. Aware of this, the Venezuelan government kept a close watch on the activities of the Venezuelan workers there in order to pre-empt any revolutionary attack emanating from the islands. The increased oil-refining activities brought a further influx of Venezuelan labour, and its attendant problems. The Curacaosche Petroleum Industrie My, Shell's refinery on Curacao, employed Venezuelan labour through a subcontractor who brought men over from Coro. According to H.L. Leyba, the Venezuelan Consul in 1927, there were some 1,500 Venezuelan workers in Curacao, mostly employed by the Shell refinery, with more on Aruba island. This concentration of workers also made political activity easier for exiled Venezuelan opposition groups, who were able to inculcate the workers with 'communist, subversive or revolutionary ideas in general against our government'.[198] Simón Betancourt, who arrived from Trinidad, and worked as a photographer, was among the principal agitators, and was in constant correspondence with Gómez's political opponent, Doroteo Flores.

The Venezuelan government was able to infiltrate these groups by planting their confidential agents among the workers, Carlos Malpica being in 1927 the 'government's agent among the Venezuelan workers'[199] at the refinery. Nevertheless, opposition to the Gómez regime was growing, and two opposition newspapers, the *Obrero Libre* and *Venezuela Libre*, were being published. Both Consul Leyba and General Argenis Azuaje, the President of Falcón, recommended in late 1927 that emigration from Venezuela to the islands should be curtailed.[200] As Pérez Soto also pointed out to Gómez, it was too easy for undesirables to enter the country by first travelling from Curacao to Amuay, in Falcón, and then by local oil tankers to Cabimas, Lagunillas, etc.[201] Gómez heeded this advice and on

12 June recommended to Azuaje that emigration from Falcón be stopped immediately.[202]

The continuous threat to the stability and peace of western Venezuela posed by the close proximity of the Dutch West Indies can be seen from the following cases. On 24 July 1928 Colonel José María Fossi, until then a trusted Gómez aide, and the ubiquitous Rafael Simón Urbina launched a revolution from La Vela de Coro (which they seized for a few hours), but which was aborted a few hours later when Azuaje's troops defeated the revolutionaries. The plan called for the revolutionaries to be reinforced by 300 Venezuelan and 90 Dominican rebels employed in Curacao, but due to the speedy collapse of the revolution these were not used. Instead, when the Dutch colonial authorities arrested Fossi and Urbina for illegally entering Aruba, a huge demonstration, which included the 300 Venezuelan and 90 Dominican workers, was organized to protest against their imprisonment.[203] A direct result of the taking of La Vela de Coro was that Pérez Soto reorganized his small armoury in readiness for an attack.[204] The attack also had the effect of tightening up security in Curacao: for instance, Domingo S. Lovera, head of the Oficina de Representación de Obreros Venezolanos at the Curacao refinery (financed by the refinery and by General Argenis Azuaje) was discovered to be an enemy of Gómez, and one of the main instigators of the protest march against the gaoling of Fossi and Urbina.[205] Leyba also secured in October 1928 the expulsion of Raúl Fernández, a leader of the 200-strong Dominican contingent on the island, because his disruptive 'presence was not conducive to good industrial relations'.[206] But despite Leyba's and Azuaje's efforts, some 90 Venezuelan workers arrived in October 1928 to work at the refinery. Leyba was of the opinion that the curtailment of emigration to the islands should be legally enforced because the 'larger the number of Venezuelan workers on the island the greater the threat to security'.[207] Industrial unrest continued on the island, and in early 1929 the oil workers protested against the murder of Hilario Montenegro, the Secretary-General of the local branch of the Communist Party.[208] Criticism of Gómez and the colonial authorities intensified during the ensuing months,[209] and an alarmed Leyba visited the Governor and the Military Commander of Curacao to draw their attention to the need for energetic measures to be taken 'to counteract the insidious campaign, while at the same time I brought to their attention the contents of a pamphlet and other inflammatory publications, from which it could be clearly deduced that a plot against our government is being organized in Curacao'.[210] This was being organized by the Partido Revolucionario Venezolano, a left-wing organization led by Carlos León with headquarters in Mexico City. Under the leadership of Gustavo Machado and Rafael Simón Urbina, a group of 80 men attacked and overpowered the 24 men guarding the island's fort and arsenal, removing three hundred rifles, a large number of automatic pistols, and three Lewis guns.[211] The SS *Maracaibo*, belonging to Shell, was commandeered and the rebels landed at La Vela de Coro where their first combat took place, and after several skirmishes at Mucuquita, Socopero and San Francisco, the group disbanded, each rebel making his own way out of the country.[212]

CONFLICT BETWEEN THE OIL INDUSTRY AND THE LOCAL POPULATION

Despite Pérez Soto's initial success at bringing law and order to Zulia, the conflict of interests between the oil companies and the local communities still remained. On 15 June 1928 what had been previously feared took place when an immense fire engulfed that part of Lagunillas which was built over the lake. The official number of houses destroyed was put at 335, but unofficial observers estimated that the true figure was above 500; material losses were calculated at over Bs.3 million. The rest of the town did not suffer because of the prevailing winds, which stopped the fire from spreading, and because of the rapid intervention of the authorities, and the 'help of the oil companies which had their drilling rigs, storage tanks, personnel and offices threatened by the fire'.[213] Most of the population fled to the surrounding towns of Tasajeras, Las Morochas, Los Riteros, and Maracaibo. The drilling operations were immediately stopped, and the oil companies started to organize a rescue operation. In Maracaibo a Junta Central de Socorros para el Incendio de Lagunillas, and a Junta de Socorros de Lagunillas were established to gather funds for the homeless,[214] and by September 1928 Bs.96,783.11 had been collected.[215] Although the inhabitants wanted to start reconstructing their town,[216] both the Venezuelan Gulf Oil Co. and Lago seized the opportunity to try to relocate the town on a new site away from the oilfield on which Lagunillas was situated.[217] Nevertheless, Pérez Soto rigorously enforced the Resolution he had enacted the previous year which defined the town's limits and site. The Concejo Municipal of Bolívar District, after receiving Arcaya's approval, enacted a further Resolution on 10 July calling for the reconstruction of the town, as well as the redefinition of its location in the same place.[218] Great resentment towards the oil companies continued, and when it was established that the fire had been started by a spark from one of the open-hearth fires igniting the oil spilt by a leaky drill near the town, the people of Lagunillas brought a claim against Lago, CPC, and Gulf Oil for damages caused by the fire. Pérez Soto informed Gómez that he had had to calm feelings because otherwise the population would have attacked the 'foreigners and their drilling rigs, presenting us with a serious dispute'.[219] This spurred other towns to take up their grievances against the companies. The people of Cabimas, for instance, on 1 August sent a telegram to Pérez Soto complaining that VOC was operating a rig less than 100 metres from the town, and enquiring whether they would be compensated for any loss incurred in the event of an accident at a well.[220] Aware that feelings ran high against the companies, Pérez Soto took up the matter immediately with VOC's local manager, who promised that it would be willing to consider bona fide claims for personal injury and damages.[221] Soon afterwards tragedy struck at Cabimas when on 17 October 1928 Earl B. Byard, an American employee of Gulf Oil, while clearing the ground for the construction of a new refinery for the company, accidentally fractured an oil pipeline belonging to VOC with his tractor. The oil spill drifted into Gulf Oil's camp for Venezuelan workers, where it ignited and caused the death of seven people.[222] Isaac Gómez, the Fiscal de Hidrocarburos, investigated the fire and concluded that the companies had been negligent about taking the necessary precautions to prevent such accidents.[223]

These fires were not the exception as many went unreported. José Jesús Paris, the Inspector y Fiscal de Minas de Hidrocarburos, during his annual visit to Cabimas and Lagunillas, revealed that a number of large fires had taken place: for example, near Cabimas, a well belonging to Gulf Oil had exploded severely burning 36 men (two Americans and the rest Venezuelans), of whom 14 later died; and another well belonging to the same company exploded near La Rosa, which remained covered in dense fog for six hours.[224] There was no medication to treat the wounded and the compensation offered by the company was derisory. Paris also drew attention to the fire at Cabimas caused by the fractured pipeline, which according to the law should have been at least three feet below ground level, and not barely below the surface. Paris' report painted a horrific picture of the whole development of the oilfields, and the effects this was having on the local population. According to him, in order to obtain valuable lands, the companies 'resorted to subterfuge and threats with the owners, obtaining the land at the price they deemed fair, and having acquired them caused a great deal of harm to the adjacent properties because they plummeted in value as nobody was willing to buy them due to the risk of fire'.[225] The towns were threatened by gas escapes, and by pipelines being too near the surface. In Cabimas, VOC had over 100 drilling rigs in the town; in addition the companies had situated their oil storage tank farms in the middle of the town, with little protection against lightning (having no lightning conductors), and, said Paris, in the case of fire: 'I consider the ring of sand surrounding the storage tanks insufficient to contain the 80,000 barrels which can accumulate there, thus allowing the fire to spread to other tanks.'[226] The entries from the lake to La Rosa, La Salina, and Lagunillas were blocked to nagivation by large numbers of drilling rigs, pumping stations, pipelines, connecting bridges, etc.;[227] moreover, the intense activities of the oil companies had covered the lake with a six-inch thick sludge of oil which contaminated the water. The treatment given to local staff left much to be desired, and the accommodation offered by the companies was poor and unhealthy.[228] The companies were also not providing the necessary hospital treatment as required by law.

THE GOVERNMENT RESPONDS

The central government was trying to rectify this situation, and in 1928 enacted a Labour Law which was welcomed by the working community of Zulia as a step in the right direction in the protection of their rights against the oil companies. The law established stricter and safer working conditions for workers, and also created the post of Comisionado Especial del Ministerio de Relaciones Interiores ante las Compañías Petroleras to supervise its implementation by the companies.[229] In addition, other government departments took a keener interest in the workers' health. For instance, the Oficina de Sanidad Nacional pointed out to the Ministry of Development the poor health service given by the Creole Oil Co. of Venezuela to its employees.[230] In addition, the new Constitution of 1929 incorporated a new clause (article 18) which prohibited the sale of communal land.[231]

In spite of these positive steps, the government had comparatively little effect on the improvement of working conditions for Venezuelan oil workers. On 3 October 1930 Jonás González Miranda was appointed Special Commissioner to the Oil Companies to inspect the living conditions of the oil camps situated in Sucre and Monagas,[232] and found that at Quiriquire, the largest oil camp, Venezuelan workers slept 'in overcrowded storage sheds',[233] which were 'inevitably infested with mosquitos'.[234] Luis Nava, another Special Commissioner to the Oil Companies, also reported on the living conditions offered by the oil companies in Zulia and Falcón, and found that several used a standard employment contract which freed them from any future claims against the companies.[235] The living quarters provided by VOC at Pueblo Nuevo, Pueblo Aparte, Cuarenta Casas, Villa Nueva, Lagunillas, Las Delicias, Campo Central Altagracia, and Campo Tasajeras were good, but the company continued to share its hospital with Gulf Oil and Lago. The situation was different at the camps at La Concepción, La Paz and Totumo in which accommodation was 'in the main old, and in a poor state of disrepair (in some cases almost in ruins), without the required sanitary facilities, and their refurbishment is absolutely indispensable'.[236] There was also no medical assistance. The accommodation furnished by Lago and Gulf Oil in the various fields in which they operated was similarly bad. The latter company housed its drilling crews working at Lagunillas and Las Morochas in six boats dotted along the lake shore. The 200 men who lived and slept here were 'extremely cramped, one on top of the other, without sanitation. Under these houses there is a thick oil slick where all the rubbish and sewage of these wretched men ends up, presenting a most repugnant sight.'[237] These conditions contrasted with the accommodation and facilities offered to foreign staff which were always better. Nava concluded:

> Studying the life that the oil workers lead it can be noted that wages are low in relation to the high cost of living, and the hard work they perform; with the added burden of the hot and generally unhealthy climate. Many complain of the ill-treatment received from the companies' foremen, too afraid to complain because of fear of losing their jobs; others find that the Employment Law is deficient and that it affords them very little protection. It can be seen that there is a great deal of discontentment among the workers, which encourages disputes and strikes.[238]

The complaints against the companies continued. In 1929 Gulf Oil erected a pipeline which crossed the entrance to Cabimas making it impossible for boats and canoes to pass.[239] The company agreed to submerge the pipeline after Pérez Soto made representations.[240] VOC, in a clear breach of its 1926 contract with the Bolívar District, refused to grant land to the people of Cabimas on which to build their houses; moreover, the company refused to allow people to build houses on their own property, and had even fenced off land for its own use which had been previously occupied by residents of Cabimas.[241] These complaints were deemed groundless by the company's legal department because the government had transferred all communal land in the district to the company.[242] In an appeal to Torres the people of Cabimas declared that the October 1926 contract with the company guaranteed them land for house construction. At Torres' request the company relented, and made more land available to the people of Cabimas.[243] But the tension between the company and the community did not diminish as the company refused to allow further development of a piece of land known as El Cobito. This was particularly galling to the inhabitants of Cabimas since with the slow-down of the oil boom many were turning to the land as a means of subsistence during the depression years, and the company in this instance was preventing them from doing so. In January 1931 a petition complaining about the ill-treatment afforded them by the company was signed by over 80 people from Cabimas, and delivered to the Inspector Técnico de Hidrocarburos.[244] Although Torres investigated the problem,[245] it was forgotten when Cayama Martínez took over as Minister of Development soon afterwards.

The inhabitants of Lagunillas fared better than Cabimas in their fight for the restitution of their communal land. As a result of Bartolomé Osorio Quintero's denouncement[246] (that, of the 15,000 hectares acquired by VOC from Samuel Meléndez and Rubén Araujo (known as Pueblo Viejo in Lagunillas), only 2,000 hectares rightfully belonged to the aforementioned people, the rest being common land), the government initiated legal proceedings against the company on 3 May 1930 at the Juzgado de Primera Instancia de Maracaibo, which on 1 July 1933 upheld the government's view that the lands should be returned to Lagunillas.[247]

Unlike the country's previous mining boom in Guayana during the latter half of the nineteenth century, the oil industry was centred and grew in the country's most prosperous and wealthy state. The result was that the industry had a greater overall disruptive effect on the economy of the state and on the well-being of its population than had been the case with the earlier mining boom.

5

Greater control of the oil industry

The rapid development of the oil industry after 1922 surprised most observers; it also posed a difficult problem for the government, which aspired to maintain control and supervision of the industry. Proper legislation covering all aspects of its activities would inevitably lag behind; furthermore, there was little point in repeating the errors of 1920 and 1921, when legislation was hastily drafted and ambiguous, as this would lead to disputes and to a possible curtailment of the companies' activities. In addition, though the country had experienced a lesser mining boom in the latter half of the nineteenth century, Gómez's government was handicapped by a dearth of qualified and honest personnel at all levels. The oil companies were able at first to act with impunity in the oil districts. In many cases they were able to act in this manner because no laws existed under which they could be controlled or their activities prosecuted. An example of this was the pollution of Lake Maracaibo. The treatment of oil workers was another area in which there was scant legislation. During these early days the government also exercised very little supervision of the exact amount of oil produced by the companies, although the companies' output determined the amount of oil revenues received by the government. Up to the enactment of the 1930 Oil Regulations, the government accepted the companies' own production figures as true, ignoring the obvious temptation that such a situation gave to the companies to adjust their figures before submitting them.

FIRST STEPS

As the industry consolidated its position during the 1920s, so too did the government, within its limitations, increase its

159

controls and supervision. The first steps were hesitant and timid so as not to frighten the oil companies, but great momentum was gathered during the mid-1920s culminating in the 1930 Oil Regulations. Although this supervision was of a tenuous nature (the effects of the 1928 Labour Law have already been discussed), it provided the framework on which future vigorous government control could be exercised. The difficulty was that during the depression years Gómez needed the companies more than the companies needed the country's oil resources, and so on a number of occasions he was forced by discretion to back away from a full confrontation with them. Nevertheless, the surprising aspect of the Gómez government was that, given its administrative limitations and the vested interests of the *gomecista* concessionaires, it reacted quickly to the impact of the industry, and took prompt action to rectify some of the more blatant abuses perpetrated by the oil companies. As the industry grew reports reached Gómez of the pressing importance of maintaining close supervision of the companies' activities.

Gómez exuded confidence in the future development of the country's oil industry, and highlighted in his Annual Message for 1923 Venezuela's new-found wealth and position as a major oil producer. However, there were some who feared that this development and the influx of foreign capital might be the cause of future foreign intervention. Gómez was able to allay these fears because the industry was composed of more than one nationality, *viz*, Anglo-Dutch and American, a situation which by its very nature would prevent any single country from intervening.

In order to supervise the exploration and exploitation of oil concessions, to oversee that any machinery and equipment imported free of duty was used by the importers, and to keep the government informed on the general development of the industry, the office of the Inspector Fiscal de Hidrocarburos, headed by Francisco M. Braschi, was created on 9 October 1923. At the same time L.F. Calvani, head of the Inspectoría de Minas and responsible, *inter alia*, for the oil industry, suggested a comparative study of the Oklahoma and California oilfields and those of Zulia and Falcón, in order to determine whether any changes needed to be made for the implementation of a

more efficient method of revenue collection.[1] Calvani did not
travel to the U.S., but instead inspected the Zulia and Falcón
oilfields, and reported to Gómez in August 1924. Calvani drew
a deplorable picture of the conditions of the oilfields, pointing
out that the lakeshore community were unable to obtain fresh
water due to oil pollution. The result was that Gómez
demanded that the oil companies supply 'water from their own
tanks, and . . . [that they should] take into account the needs of
the humble people living on the lake shore'.[2]

In order to attract and encourage American investors to
Venezuela, Lucio Baldó, recently appointed the Director of
CVP, attended in October 1923 the International Oil Exhibi-
tion at Tulsa, Oklahoma, representing Venezuela, and in
August 1924 Gulf Oil found oil under Lake Maracaibo,
confirming the view that the country was a large producer of
oil. In addition, the price of Venezuelan crude delivered at New
York was the lowest of any oil produced which meant that
Venezuela's oil in the future would be 'of extraordinary
importance'.[3] A further impetus to Venezuela's development
was provided by fears for Mexican deposits.[4] Nevertheless,
there were still several important difficulties to be resolved by
both the government and the companies. One of the most
pressing was that created by land demarcation disputes which
had resulted from hastily drawn-up contracts which in some
cases partially fixed borders on the undetermined state
boundaries of Zulia, Falcón, and Mérida. The problem was
further intensified by the backlog of survey plans which had
flooded the Ministry in 1921, and which still needed to be
checked and corrected.[5]

Prices had slumped dramatically in 1925 on account of the
great number of oil concessions on the market. For example,
600,000 hectares in Monagas State were sold for Bs.3 per
hectare, and in Zamora State oil concessions were fetching
Bs.1.50 per hectare.[6] The government, too, had not received
any substantial increase in oil revenues since 1922 because most
of the companies entering the country after the First World War
were still in the exploration stage. This did not apply, however,
to Shell, and the government once again took a keen interest in
the ambiguous Aranguren oil title held by VOC for the purpose
of extracting additional revenue from the company, and

threatened to annul the concession. After some hasty negotiations the company agreed to pay Bs.10 million,[7] in return for which it obtained an extension to its exploration period from five to seven years (starting from 25 April 1921). It also renounced the agreement to a decrease in exploitation tax from Bs.2 to Bs.1, paying now Bs.2 per ton of oil produced. Finally, no further doubts about the validity of its concession were to remain. Both the company and the government were happy with the outcome, for the company retained its valuable concession intact, and the government obtained a Bs.10 million injection in oil revenues; but most important, the company (which was the largest oil producer in the Shell group) agreed to a 100 per cent tax increase on oil produced, an important achievement on account of the great size of the company's production.

On 24 June 1925 a new Constitution was sanctioned by Congress which, *inter alia*, declared that mining and oil concessions were no longer subject to congressional approval, leaving sole control over this matter to the Executive. Consequently, a new Oil Law was enacted on 18 July 1925 to harmonize the law governing the oil industry with the new Constitution. Some of the 'obscure provisions'[8] contained in the previous Oil Laws were also changed, to the companies' delight. Minor changes included the provision that oil pipelines under the sea, lakes or rivers should not block shipping; that the companies would be allowed the use of wood found on communal land for their drilling operations only; and that the details to be included in companies' annual reports submitted to the Ministry were to be outlined more clearly.

The refinery and transhipment facilities which had been built in the Dutch West Indies were a constant source of annoyance to Gómez and his government, for they saw the Dutch colony benefiting from Venezuela's oil resources. In addition, such facilities made it difficult for the goverment to check the export figures given by the companies. The government was therefore attracted to the plan proposed by William F. Buckley, the head of Pantepec Consolidated Co. of Venezuela, to build a major deep-water port on the Paraguaná Peninsula (Falcón State) which would be the only authorized transhipment port for oil exports. By centralizing the transhipment activities in one port the government would also be able to check the production figures given by the oil companies.

BUCKLEY'S OIL TRANSHIPMENT PORT

In 1924 Buckley presented to the Ministries of Development, Finance, and Public Works his project to build a commercial and oil transhipment port at Salinas Bay. The Compañía Marítima Paraguaná C.A., which would build the port, was registered in Caracas on 22 May 1924, and one month later, on 24 June, it was granted permission to carry out the project. The companies which had obtained permission to build transhipment ports on Paraguaná objected strongly to this new project because, in addition to making their properties valueless, the oil companies operating in Venezuela would, under Buckley's contract, have to ship their oil through the new port, and pay, on top of the usual port dues, a new tax on every ton shipped. This new tax they considered illegal because it had not been included in their initial operating contracts.[9] The government, for its part, was attracted to this project for the following reasons: first, the need for a transhipment port for merchandise to and from Maracaibo to eliminate Curacao; second, because oil produced in the country would be transhipped from a Venezuelan port, and not from the Dutch colony; third, because Salinas Bay offered the only suitable deep-water bay in the Paraguaná Peninsula where storage tanks and facilities could be built; and, finally, because the government wanted to give the

oil industry the security that in the future it would not be subject to other contributions which had not been stipulated in the Oil Law, and to prevent a repetition of what occurred in México, where taxes increased in 1924 to thirty seven American gold cents, thus ensuring the stagnation of the mining industry.[10]

Buckley proposed that the port cover the area between Salinas Point and Los Taques Point, on land owned by his own company, but at the last meeting with Antonio Alamo before signing the contract, the Minister insisted that the port's jurisdiction be extended from San Román Cape to Oriboro Point, with the result that it now covered the other companies' terminals (see Map).

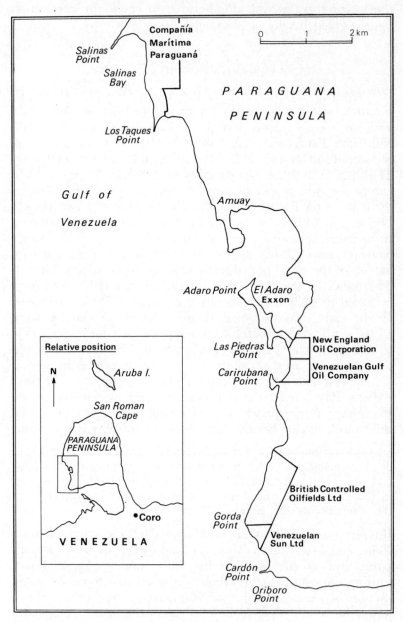

Location of transhipment oil terminals on the Paraguaná Peninsula

Buckley's contract caused 'untold consternation among American and British interests',[11] which were quick to react to get the contract rescinded. Under the leadership of Rutherford Bingham of Exxon, an emergency meeting was called of all the companies affected to press the government to change its mind. A delegation composed of Rutherford Bingham, Fred H. Kay, Paul Adams, and Alejandro Pietri[12] went to Maracay on 28 June 1924 to confer with Gómez over this matter, and to inform him of their objections. Gómez, however, refused to meet them and suggested through an aide that Antonio Alamo be consulted on the matter. Due to the companies' opposition, and that of the Dutch West Indies, Congress felt it expedient to postpone a decision on the project. Buckley welcomed this decision for the government would be able to study the project in greater detail, and so determine more fully its advantages.[13]

In early July the companies signed two petitions, one asking for the opening of the Maracaibo sandbar, and another requesting the government to establish a customs house on the western coast of Paraguaná because, they alleged, it was difficult to conduct business at the nearest customs house of La Vela de Coro.[14] Buckley interpreted these requests as a means for the companies to gain time, the better to prevent his contract from being adopted. Nevertheless, Buckley suggested that the government investigate the possibilities of his plan, and, in addition, offered to comply with any amendments which the government might propose. Gómez agreed to this and on 31 October 1924 appointed Manuel Cipriano Pérez to study the feasibility of the Salinas port, and to discover which bay between Salinas Point and Cardón Point on Paraguaná was the most suitable for a deep-water port.[15]

In the meantime Buckley, assisted by Arcaya, the Venezuelan Minister in Washington, who held large tracts of land in Paraguaná, redrafted the contract, which in the spring of 1925 he presented to the government under the name of Henry Hinds.[16] However, Exxon informed the government that Buckley had been deported from Mexico, and was considered *persona non grata* by the U.S. government, with the result that approval of the project was further delayed while these allegations were investigated by Alamo. The allegations were proved groundless, but Buckley faced further opposition when Arcaya, who up to then had supported the project, changed sides and joined Exxon. Arcaya, now the Minister of the Interior, raised among his Cabinet colleagues doubts about the legal, economic and diplomatic character of the project. For instance, he was sure that Gulf Oil, through the American Treasury Secretary Mellon (whose family owned the company), and the American Vice-President General Charles G. Dawes (a director of Pure Oil Co.), would force the U.S. government to oppose the contract.[17] Despite opposition, the Cabinet discussed and approved the port in November 1925, with the exception of Arcaya who requested that the final decision be postponed for the preposterous reason 'that he had been unable to study the contract'.[18]

Buckley, besides redrafting the contract, also tried to put pressure on the government to approve it. In December 1925 an article inspired by him appeared under the by-line of Thomas F. Lee in which it was stated that a Zulian secessionist movement, backed by the oil companies, was ready to form a Republic of Zulia. The support for such action was small but growing stronger with the oil wealth generated in the state.[19] Consequently, if the sandbar that blocked the entrance to Lake Maracaibo was dredged allowing ocean-going tankers to enter, then the state's wealth would increase even further, stimulating more support for the secessionists' cause. For political reasons, therefore, the Federal government had no intention of opening the sandbar, and the oil companies were unwilling to finance the venture themselves, preferring to establish individual terminal ports on the Paraguaná Peninsula. 'Lee' also argued that one port at Salinas Bay would be better able to handle all the oil produced than would a number of individual terminals. For example, in 1915 six terminal stations were built at Lobos Island, off Tampico, within a short distance of each other, and at an estimated cost of $100 million. This was very inefficient when one large port could adequately handle all the oil, and in addition would be less expensive to all concerned.[20]

Although the article by 'Lee' raised alarm over the situation in Zulia, it did not force the government into action. In early 1926, however, Manuel Cipriano Pérez's report recommended Salinas Bay as the most suitable place between Salinas Point and Cardón Point to establish a port, and on 3 April 1926 the government decreed the establishment of an oil transhipment port there in west Paraguaná.[21] The question now was to determine the site of the port. The five companies which had previously opposed the establishment of such a port, now welcomed the scheme but suggested that it be situated not at Salinas Bay, but between Cardón Point and Adaro Point.[22] Consequently, in May 1926, the port was discussed by the Cabinet which decided that a commission, composed of Dr Centeno Grau, the Minister of Finance, Arcaya, the Minister of the Interior, Dr Tomás Bueno, the Minister of Public Works and representatives of the five companies opposed to the siting of the port at Salinas Bay should proceed to study and recommend the 'best site for the port, as the wish of the President is to conciliate the interests of the government and the oil companies'.[23]

Much to Buckley's surprise, and to the consternation of the oil companies, on 24 July 1926 the government decreed Salinas Bay as the site for the oil port. The Dutch government, Shell and the American oil companies immediately started to exert their influence on the government to have the decree annulled. Although they recognized the government's right to decree the construction of such a port, they also insisted on their contractual right to build transhipment terminals wherever they wished. It was also becoming apparent that the building of a port at Salinas Bay would inevitably lead to a confrontation between the government and the oil companies, especially since the government looked with disfavour on the construction of refineries and transhipment terminals in Dutch territory. There were also intimations that the Venezuelan government would in the future 'discriminate in some way in favour of those companies which maintained their refineries and terminals in Venezuela'.[24] Nevertheless, it had become increasingly evident to the government that the oil companies would not accept the Salinas Bay project. Therefore, in order to maintain good relations between the oil industry in general and the government (the most important priority for Gómez, especially at a time when it appeared that there was a decline in the activities of the oil companies), the annulment of the Salinas Bay decree would be necessary. At the same time a new Minister of Public Works, Dr José Ignacio Cárdenas, was appointed who soon expressed 'a desire to rescind the Decree of July 24, 1926'[25] because of his close contact with Dutch interests and with Shell. At a Cabinet meeting in October Dr Cárdenas brought up the Salinas Bay project, stating that the port was unsuitable and that the oil companies were opposed to it. There was no further discussion by the Cabinet, and when it was proposed to rescind the decree establishing the port of Salinas, no 'objection was raised by the other Ministers, though it is known that a majority of them favoured Salinas'.[26] Consequently, to the oil companies' jubilation, on 26 September 1927 the decree revoking the establishment of the port of Salinas was published after a 'subsequent investigation revealed that the bay was not suitable for that purpose'.[27] There was still a need to select the site for an oil transhipment port as established in the decree of 3 April 1926, but it was expected that the oil companies would be able to choose their own sites, and, as Arcaya hinted in his Annual Report for 1926, the government could develop and finance its own port.

A NATIONAL OIL REFINERY AND THE TURIAMO PORT PROJECT

Although Buckley's plan for an oil transhipment port was shelved there was still interest among government circles in building a large commercial port and refinery to handle the government's growing oil royalties which would diminish the important role played by the Dutch West Indies in Venezuela's trade. Some circles believed that the government could achieve a greater return from its oil royalties if the oil was refined in the country and sold on the international market. According to González Rincones and Lucio Baldó, CVP engaged the services of Dr Roy S. Macelwee (Director of the U.S. Office of Internal and External Commerce) to determine the most suitable site for such a port. The bay of Turiamo was chosen, and Bethlehem Steel Co. was to build a refinery with the financial backing of the Morgan Guaranty Trust Co., Hemphill Noyes and Co., J. Henry Schroder Banking Corp., Dillon Read and Co., and J. Jackson Storer and Co.[28]

    The large annual accumulation of oil stocks accruing to the government also attracted private entrepreneurs ready to seek a government contract that would grant them refining and marketing rights over the oil. One such entrepreneur was Antonio Andújar, head of the Andes Petroleum Corp., a Spaniard resident in New York who in 1926, with his Venezuelan backers Alfredo Jahn, Félix Galavis, Pius Schlangeter, and Pedro José Rojas, proposed the formation of a vertically integrated oil company (with joint foreign and Venezuelan capital) which would exploit the country's oil resources (as well as the government's oil royalty), establishing a refinery at Turiamo and marketing the refined products both locally and internationally. Andújar believed that such a scheme would create greater revenue for the Venezuelan government from the profits arising out of the venture than it had achieved by selling its oil royalties to the companies. It would provide cheaper oil for internal consumption, and would stimulate national capital to further invest in the oil industry, and, finally, it would allow Venezuela to diversify its oil markets, for the company could enter into supply contracts with the newly created state oil monopolies of Spain and France.[29]

The scheme was formally submitted to the government in early January 1927, and was well received initially, for Andújar left for Europe in March to secure oil supply contracts. After spending some time in New York, Andújar arrived in Spain in June, and soon secured a contract with a strong banking group, headed by Banco Urquijo, that had been awarded the state oil monopoly. Soon afterwards a Sociedad Venezolana Española was formed.[30]

However, although Gómez had assured Andújar that approval of his scheme was only a formality, once the details of the project were known it was rejected by the Cabinet.[31] There were fears that Andújar would be unable to raise the necessary financing from local sources, and would be forced to seek outside help, thus losing control of the project. The Venezuelan government would in turn forfeit control of part of its own revenue, since ownership of the government's royalties would be in the hands of foreigners. Although Andújar persisted in his venture, submitting modified versions of his project, he was unsuccessful in the end. Soon after this the government entered into an agreement to sell its 5 per cent royalty received from Lago's maritime zones 1–3 to the Monopolio de Petróleos de España, which would supply the latter's Tenerife refinery.[32]

Although Andújar's project did not prosper it encouraged others to make similar proposals. Consequently, in the ensuing years, several projects were submitted for a national refinery to handle the government's oil royalties to be established at the recently created port of Turiamo. The first of these projects followed Lecuna's advice very closely for the government itself, a few months after rejecting Andújar's scheme, proposed to establish a refinery at Turiamo.

After the Salinas Bay project was abandoned the government was anxious to find a site for a new port, and Gómez ordered Luis Vélez, then the Minister of Public Works, to commission a feasibility study on Turiamo Bay, which was carried out by the Martin Engineering Co. Gómez's intention was to create a free port area where customs duties would only be levied on goods leaving the port, resulting in the importation of more merchandise without the immediate payment of customs duties. The idea for such a port was only possible because of the large surplus of revenues accruing to the government from the oil industry, but Gómez also felt that Turiamo could become an important oil centre in its own right.[33] Gómez donated the land on which the port would be built, and by May 1928 was very excited about the enterprise. Colonel William Warfield, representing the Bethlehem Steel Co., arrived in Caracas with a scheme to build and finance an oil refinery at Turiamo. Dillon Read and Co. and the J. Henry Schroder Banking Corp. agreed to raise $12 million (the cost of the refinery) on the New York Stock Exchange under the following conditions: that the Venezuelan government would agree to guarantee the principal and interest on the bonds from its oil revenue, including surface taxes; and that the enterprise be operated in such a way as to give the government a very substantial portion of the expected earnings (80 per cent of the profits would go to the government, with the remaining 20 per cent to the originators and managers of the enterprise, or other private individuals). According to C. van H. Engert, the U.S. Chargé d'Affaires, the U.S. bankers would have liked to submit a sound non-speculative proposition based on the assumption that the Venezuelan government was anxious to realize a larger profit out of its royalty oil, but did not wish to use government funds or borrow money to finance the scheme. But as negotiations progressed it became apparent to Colonel Warfield that Gómez and his entourage 'were really not so much interested in the profit that the government could make out of refining petroleum as they were in the exorbitant profits they themselves expected to derive from the sale of the land for the refinery and from the transfer of the concession for themselves to the Company'.[34] Gómez did not want to pledge the oil revenues to guarantee the principal and interest on the bonds, although during negotiations he intimated that the government 'might possibly consider giving some kind of guarantee which would become effective if for any reason the supply of royalty oil should fail'.[35] Colonel Warfield left for New York on 15 May to explain the position of the Venezuelan government to the New York bankers, who, as expected, withdrew their support.

However, the plans for Turiamo continued. The Martin Engineering Co. reported in November that Turiamo was 'an ideal site for the construction of a duty free port, and its accompaning infrastructure'.[36] The bay was deep and well protected on all three sides by high mountains. The advantages of Turiamo were apparent to all concerned. It was also true that whoever secured such a contract would be able to reap large profits. Torres, Gómez's most competent Minister of Development, felt that the establishment of such a refinery would undoubtedly reap great financial rewards for the government, but that the country as a whole would not benefit because agriculture would suffer from a lack of labour, so that the net effect would be 'a rich government and poor people'.[37] In addition, he did not feel that the government would be able to exploit successfully its royalty because of a lack of experts, and because of the national attitude that 'whatever belongs to the government is for everybody to share'.[38] Over the remaining years of Gómez's rule many other projects were submitted to the government for a national refinery (see below), all stressing the advantages of this to the government and to the nation.

THE GOVERNMENT EXERCISES GREATER CONTROL OVER THE OIL INDUSTRY

There was a pressing need to bring order to the industry as a result of the rapid and uncontrolled increase in oil activities in the country. The Oil Laws had always contained provisions to safeguard the rights and safety of people living near the centres of oil activity, and also the interests of the government in general. Consequently, the companies were obliged to take reasonable precautions to prevent fires and to avoid unnecessary losses to the government by improper drilling and the abandonment of wells. The companies had little regard for the safety of the local population, and in some instances developed 'line fights' which were detrimental for the long-term development of the industry. Thus on 26 January 1927 Alamo, the Minister of Development, issued Resolution No. 49, designed to prevent companies drilling too near to each other, to minimize fire risks, and to check conflicts between competing companies. The resolution forced the companies to space their wells 75 metres apart and prevented them from drilling within 37.5 metres of the boundaries of their concessions. Although most companies welcomed the resolution, they felt that this was only the beginning of a series of government regulations which would eventually restrict their freedom of operation.

The continuous reports reaching Gómez on oil pollution affecting Zulia meant that by the end of 1927 the problem could neither be ignored nor left any longer in the complacent hands of the oil companies. As an initial step the government appointed a commission, composed of Dr Manuel Toledo Rojas, Diego Bautista Urbaneja, and Siro Vásquez, to investigate the extent of the problem on Lake Maracaibo. The commission, which arrived in Maracaibo in January 1928, visited Lagunillas, La Rosa, Cabimas, Ambrosio, La Misión, El Mene, La Rita, and Bariancas, and found that 'the waters were full of oil, and the lake shore also polluted by an extensive and thick layer of that substance, and the residents of the settlements complain about it because they are unable to use the water for domestic purposes'.[39] The main cause of this contamination was the companies' careless operational procedures. When oil was transferred from a well over the lake it was taken to special tanks where the mud, rock, sand, and other debris were allowed to settle. The oil would then be drained off and the oil-saturated mud would be thrown into the lake. Some of the 'gushers' that came in over the lake were difficult to control; for example, Lago's well No. 444 exploded and spilled oil over the lake for a week before it was brought under control. Oil was also lost over the lake when the loading pipe was disconnected from the tankers; and when these returned empty from the Dutch West Indies more oil entered the lake when their ballast of water and residual oil was discharged. Finally, workers were negligent about filling the storage tanks, allowing oil to spill over and drift on the lake. Alexander K. Sloan, the U.S. Consul at Maracaibo, considered the commission's report had been 'very mild in its statements as to the effects of the presence of oil in the waters of Lake Maracaibo'.[40] The waters of the lake were 'so brackish, except in the southern section, that irrigation ditches through which lake waters are forced are covered with a salt deposit'.[41]

As a result of the commission's report, Congress enacted on 13 July 1928 a Ley de Vigilancia para impedir la contaminación de las aguas por el petróleo. Willis C. Cook, the American Minister, remarked that oil companies judged the law to be 'fair and reasonable',[42] and it appears to have done some good as Alamo declared in April 1929 that oil spillages had decreased considerably.[43] Other less serious problems were firmly dealt with by the government. For instance, in December 1927, it rejected the survey plans presented by Lago because the company wanted to group together in a central block in the middle of the lake the National Reserves it had to return to the government. Such an arrangement meant that the National Reserves would lose some of their attraction as drilling would be difficult at the deep end of the lake. The government requested that the National Reserves adjoin the blocks retained by the company, to which it eventually agreed.[44] In other aspects the government was more lenient, allowing, for example, the companies in 1928 to employ foreign doctors to practise medicine at their own hospitals.[45]

THE 1928 OIL LAW

On 18 July 1928 another Oil Law was enacted which did not differ materially from that of 1925. Of the minor changes, the most important was the abolition of the 2.5 per cent special tax levied on gross income accruing to refiners who did not possess their own source of oil.

This same year also experienced a revival of the oil companies' activities after the previous year's temporary lull. A total of 338 new concessions were awarded during 1928, and a production of 303,000 barrels per day in July set a new monthly record.[46] This optimistic feeling was soon dispelled, however, with the advent of the Depression, and the continual overproduction of oil in the U.S. The major oil companies operating in Venezuela, following the spirit of the 1928 Pool Agreement, decided to limit production in the country to 120 million barrels per annum.[47] Consequently, 1929 was the first year since 1920 in which production did not double because the companies, as Torres was fully cognizant, had to 'place the oil within certain limits to avoid the plethora of products on the market, and by the same token, the fall in prices'.[48]

Meanwhile, the government was preparing to impose stricter controls over the industry. In mid-1927 L.F. Calvani, head of the Sala Técnica de Minas, travelled to the U.S. to compare the American oil industry with that of Venezuela. He was very much impressed with the laws and measures adopted by California to prevent oil spillage, and believed that 'if tomorrow the same or similar legislation was applied to American companies operating in the country, they would be unable to oppose it'.[49] Calvani also attended the 6th International Oil Fair and Congress at Tulsa, Oklahoma, where he was given preferential treatment, and was also involved in discussions over a proposal to limit foreign oil imports into the U.S., which ended without reaching a firm agreement.[50]

TORRES' SECOND TERM AS MINISTER OF DEVELOPMENT

Despite the overwhelming encouragement of Congress, Gómez, at the end of his seven-year term of office as President in 1929, did not want to remain as the head of the government. However, he was determined to hold on to the reins of power, and consequently Congress reformed the Constitution, creating for Gómez the new autonomous post of Commander-in-Chief of the country's armed forces. This enabled Gómez, together with the titular head of the government, to 'appoint and dismiss ministers, govern the Federal District, convene Congress to extraordinary sessions, suspend constitutional guarantees, command the armed forces, and grant pardons'.[51] In his new capacity he returned to Maracay and Juan Bautista Pérez, the former head of the Federal Court of Cassation, was elected President. A few weeks later Gumersindo Torres was appointed the Minister of Development for the second time. This was interpreted by the companies as a sign of the tougher new direction the government would take towards the oil industry.

Torres took up the challenge of his appointment with vigour, and began by applying more strictly the conditions of exoneration from customs duties enjoyed by the oil companies, stopping the importation of petrol, paraffin, and other oil products by the oil companies free of customs duties altogether.[52] Furthermore, after several complaints, Torres warned Exxon that it had to provide adequate medical treatment and facilities for its employees in its camps in Monagas State, which the company agreed to do. A more serious problem was that Lago which held 1,143,341 hectares over Lake Maracaibo, had still not paid the initial exploitation tax and surface tax. The company had been unable to come to an agreement over the interpretation of article 42 of the 1925 Oil Law which ruled its concessions, thus leaving unclear the area on which to calculate its tax. After negotiations, during which Calvani made a survey of the concessions, the company agreed on 26 March 1930 to pay Bs.1 per hectare initial exploitation tax on half its acreage, a sum of Bs.541,670.60[53] and the government later received Bs.1.6 million in back taxes. Torres also increased taxes charged on National Reserves which would be awarded in the future. Ever since these had been created the government had been given the right to stipulate 'special advantages for the Nation in relation to taxation'.[54] The 1928 Oil Law was no exception, and on 19 November 1929 Torres, implementing article 40, increased taxes charged on National Reserves to the following rates: initial payment went up from Bs.2 to Bs.15 per hectare; surface tax went up from Bs.2 to Bs.4 per hectare for the first three years, Bs.4 to Bs.5 for the next 27 years, and Bs.5 to Bs.8 for the remaining 10 years of the concession; royalty payments went up from 10 per cent to 15 per cent. The U.S. Chargé d'Affaires reported his fears that this decree 'may be the entering wedge for the revision of the entire Petroleum Legislation, and it therefore deserves the closest attention of all American oil companies interested in Venezuela even though they may not be directly concerned with the future exploitation of natural reserves'.[55] The companies accepted Torres' initial changes and were not unduly alarmed when he raised taxes charged on National Reserves since they would not be applied retroactively. Nevertheless, Torres would face stronger opposition when he tried to collect higher buoy taxes from two of the companies.

In order to stimulate crude oil refining in the country, the government, encouraged by Arcaya, the Minister of the Interior, sought in 1926 to impose a buoy tax on crude oil exports from Maracaibo. Although there was immediate opposition to this idea (the exception being Buckley since the tax would not be charged on coastal trade, thus making the port at Salinas more attractive to all the companies), Congress, on 28 June 1926, enacted the first Buoy Law. However, as a result of the opposition expressed by the oil companies, the tax was reduced from Bs.2 to Bs.1 per ton on ships leaving Maracaibo, and coastal trade was exempted from paying the tax. But as predicted by Warren W. Smith of the Pantepec Oil Co. of Venezuela, the companies did not establish refineries in Venezuela and the government became increasingly resentful of the establishment of larger refineries in Aruba and Curacao. Consequently, in 1928, the government, through Senators G. Terrero Atienza, Juan E. Paris, and Juan Guerra, presented a bill to Congress calling for the buoy tax to be increased from Bs.1 to Bs.2 per ton on vessels leaving Maracaibo and from nil to Bs.1 per ton on vessels which left Maracaibo transporting agricultural exports, with refined products untaxed.[56]

Lago and Shell immediately made representations to the government, and a new draft bill was submitted 'in such form as to allow for a *50% reduction* of the Bs.2 Buoy tax for the benefit of those companies which had made their arrangements for crude oil exploitation prior to the passage of the new Statute'.[57] The new law of 2 July 1928 contained all the modifications requested by the companies, but the clause which reduced by half the tax imposed only applied to companies which were producing and exporting oil to refineries or transhipment ports when the law was enacted. In the latter case, the company, in order to benefit from the reduction, needed to demonstrate that it had also started construction on a refinery. The companies

were very concerned about the new law, and Shell and Lago
decided to pool their resources, and to act in close co-operation
to have the law abrogated. After consulting the other companies
it was decided that each one should address a petition to the
government 'pointing out that the Buoy tax law really was a
charge on crude oil and as such was not payable by the
Company concerned'.[58] This they did on 20 August 1928. The
companies' objections to the law stemmed from the belief that
although the buoy tax was a port charge, it amounted to an
exclusive tax on crude oil exports due to the exceptions and
privileges granted to ships in other trades. The companies
therefore claimed that the law did not apply to them since their
contractual obligations did not allow further taxes to be levied
on their activities, and it was also unconstitutional because
export taxes were expressly prohibited in the Constitution. The
regulations of the law were also unworkable because the 50 per
cent reduction of the tax would only be granted to companies
which had invested in transhipment and refinery facilities, and
such a condition, which presupposed that the 'operating
companies (had) their own refineries is not to be found in the
Buoy Tax Law, and the Reglamento therefore goes outside the
scope of said Law'.[59] What the law did prescribe was that at the
time of enactment the companies which were already exporting
oil to either their transhipment terminals or the Dutch West
Indies would receive the reduction. The regulations, on the
other hand, established the condition that the reduction in tax
would be subject to the companies owning refineries or transhipment terminals in the country, and, therefore, went outside
the scope of the law. The regulations also limited the reduction
of tax to a period of 8 years (article 2), which was considered by
the companies to be too short and insufficient 'to compensate
for the interest paid on capital invested in refineries and
transhipment facilities in Aruba and Curacao'.[60] The regulations also appeared to curtail oil exports, for article 3
established that the government would determine the number
of ships and the tonnage that would benefit from the tax
reduction. The companies considered this article an infringe-
ment of their freedom since the government could decide the
number of tankers and the tonnage needed by each of them in
their operations.

For the next seven months the companies' petition remained dormant, but in March 1929 the government demanded that the companies pay the full Bs.2 per ton of the tax, which the companies refused to do until the question had been satisfactorily settled. After further protests by the companies, and a report by D. Giménez,[61] the Administrator of Customs at Maracaibo, on the impatience the companies felt with the Ministry's delay, the government agreed to settle the dispute by compromise. On 12 September 1929 the first of several meetings took place between government and company lawyers, with the result that the government lawyers saw the need to modify the regulations, admitting that the companies were correct in their objections and requesting the companies to 'prepare a draft of what in their eyes would be a satisfactory Reglamento'.[62]

A few days after Torres' appointment as the Minister of Development in early October, lawyers from both sides met, and the government representatives were presented with a draft of the new regulations. W.T.S. Doyle, the head of Shell in Venezuela, noted that there were still two outstanding points which the government did not accept: that the reduction grant could not take place over the life of the concession; and the companies' insistence that the tax reduction should be retroactive to cover the time elapsed since the promulgation of the new law. In early November, following more meetings between the two sides, the problem was discussed by the Cabinet. At that meeting the Finance Minister submitted a Memorandum which stated that the oil companies did not want to recognize the buoy tax nor would they pay such taxes, and requested suggestions from the rest of the Cabinet to resolve the matter. Torres felt that the question hinged on whether 'to allow the law to be mocked or to have the conviction to force its compliance'[63] by insisting that the companies pay what they owed to the government, because the longer it took to find a solution the greater the difficulty in collecting the outstanding revenue. The Cabinet agreed with Torres, who took charge of the affair on the condition that he received the Cabinet's overwhelming support in obtaining from the oil companies all the outstanding taxes without making any 'deals, as I wanted to maintain my name above suspicion'.[64]

After the Cabinet meeting Torres informed F.C. Pannill, the

head of Lago, and Doyle that they owed him 'a great favour'[65] because he had been able to restrain the Finance Minister temporarily from striking a 'hard blow'[66] against them. The government, with Gómez's approval, was willing to take the companies to court in order to force them to pay the buoy taxes. Once the lawsuit was presented to the court the Attorney-General would place an embargo on their property, a move which would ensure for the government the payment of any fine inflicted by the court, and also of the outstanding taxes. Such an embargo would cause the companies much inconvenience and financial loss since the government was confident of winning the case. After further discussion, Torres was willing to come to an 'arrangement' with the companies by which a new *Reglamento* would be enacted authorizing the reduction grant to last 25 years, while the companies would pay in full the back taxes, amounting to Bs.12,494,000. This money was duly paid to the government on 9 December 1929.[67]

On 14 December 1929 the new *Reglamento* was published, which at first sight did not appear 'quite satisfactory on every point'[68] to the oil companies. Nevertheless, the companies believed that as Torres had acted in good faith a solution could be found with the government. At the opening of negotiations Doyle found Torres 'so reasonable that I am not without hope that we will at least be able to obtain a workable reduction grant'.[69] At their meetings several ambiguous points of the new *Reglamento* were brought to the Minister's attention. For example, article 7, which proposed setting out the maximum quantity of oil allowed to be exported at the reduced rate, was interpreted by the companies as interference by the government with a view to regulating the companies' activities. This had not been contemplated by the 1928 Oil Law. Lago indicated that 'strictly speaking it cannot comply with the requirements of section (e) and (g) of Article 4 because it does not own any refineries in the Antilles'.[70] The refinery at Aruba where the company sent its oil was owned by the Pan American Petroleum and Transport Co. which, though a subsidiary of SOC (Ind), was technically 'entirely independent of the Lago Corporation'.[71] In addition:

A question which was evidently not foreseen by the government arises in connection with the interval between November 21, 1929 i.e. date of the Regulations and December 14, 1929, the date of their publication in the Gazette. The companies fear that unless it is specifically stated that the date is [manuscript illegible] pending transactions they may be called upon for back-taxes for the intervening period too.[72]

A typographical error in the Official Gazette giving the date of the Buoy Tax Law as 9 July instead of 9 August meant that the regulations would have to be published again, giving the government the chance to incorporate the companies' suggestions. After further negotiations, in which the American legation was closely involved,[73] on 13 February 1930 a new decree was issued with a revised version of the *Reglamento* embodying the companies' recommendations, viz, there would be no limit to the amount of oil exported to the Dutch West Indies at the reduced rate; the duration of the reduction grant would be 25 years; no further back taxes would be claimed; and ownership of refineries at Aruba and Curacao was not necessary in order to benefit from the reduction. To acquire the reduction companies had to be exporting oil to be refined in the Dutch West Indies before the Buoy Tax Law of 1928 was enacted. CPC, VOC, Lago and the British Controlled Oilfields Ltd all achieved a reduction on the buoy tax on oil sent to the Dutch West Indies,[74] but CDC was refused its reduction on 10 July 1930 because it did not possess a refinery on the Dutch West Indies before the law was enacted.[75]

Although the government's intention was to encourage the oil companies to refine oil in the country, the lack of specialized technical skill in the administration made this policy unsuccessful. Soon after the reduction in buoy taxes it became known that false tonnages on company tankers were being declared, resulting in a loss in government revenues. This meant that more crude oil than had hitherto been suspected was being shipped out of the country, undermining the government's policy to increase the country's refining capacity. In many cases the tonnage registered at Maracaibo did not tally with Lloyd's Register. This discrepancy was brought to Torres' attention in a confidential Memorandum in which it was shown that, for each tanker leaving Maracaibo belonging to the Curacaosche Scheep Mj. (Shell), Lago Shipping Co. and Arend Petro.Mj., the government lost at least Bs.1,100 in buoy tax as a result of the low tonnage declared. In June 1930 these three companies alone evaded buoy taxes amounting to Bs.539,000.[76] Another way of detecting tax evasion, in addition to checking Lloyd's Register, was to examine the amount of oil exported and compare this to the buoy tax paid. Rincón did this for the period July–December 1934, and found that while the buoy tax amounted to Bs.3,733,798, the amount of oil exported was 7,035,681 tons for the same period, so that even if all the companies enjoyed the buoy tax reduction and paid only Bs.1 per ton the companies should have paid Bs.7,035,681. The government, therefore, lost a minimum of Bs.3,301,883,[77] equivalent to 12.7 per cent of the total taxes paid by the companies for the period in question. The real difference, however, was greater since not all the companies enjoyed the reduction grant, and not all crude oil exports went to the Dutch West Indies for refining.

The government was also cheated when part of the oil declared to be refined in the Dutch Antilles was instead re-exported in its crude form to the U.S. or Europe. The coastal tankers to the Paraguaná oil terminals also obtained the reduction grant. Gulf Oil, for example, had 14 tankers registered in Venezuela which shipped oil from the Maracaibo basin to Paraguaná, thus being exempted from 50 per cent of the tax,[78] but which to all intents and purposes were foreign since their captains and crews were all foreign.[79]

Some of these evasive tactics were discussed by the Cabinet, which agreed that the government should press the companies to make up the difference.[80] Villegas Pullido, the Attorney-General, was of the opinion that an error in judgement did not mean that the companies were exempted from tax payment, and advised Torres that the Ministry of Development could claim the difference owed.[81] Torres requested Leyba, the Venezuelan Consul at Curacao, to make a 'comparative list of the registered tonnage and the amount of oil carried by the same ship from Maracaibo'[82] to Aruba and Curacao. The companies, however, stated that the difference in tonnage was simply that between capacity tonnage and gross weight,[83] and consequently they had declared the net tonnage which was calculated at 45 per cent of the ship's gross tonnage effective capacity.[84] Under the 1918 Customs Law this was illegal because the companies were bound to declare total gross tonnage, but the government had allowed this discrepancy to continue in the case of merchant shipping. There was, however, no reason for allowing oil tankers to continue the practice since their storage holds were fully utilized carrying oil. Although the government was fully aware of the situation it did not press the companies to make up the difference since, with the enactment of the 1930 Oil Regulations, company opposition to the government (with the companies at one point threatening to pull out of the country) increased dramatically.

THE 1930 OIL REGULATIONS

Torres had worked hard since his reappointment as the Minister of Development to bring order to the oil industry. His biggest internal problem was dealing with the people who wanted to cash in on the country's oil wealth.[85] It was also becoming clear that his effectiveness, and that of the Ministry, were constricted by the lack both of adequate personnel and of technical information about the industry. He was amazed that 'we have no technical experts for such an important industry',[86] and in his Annual Message for 1929 proposed to organize a team of specialists who could deal on equal terms with the oil companies. It was clear that the *carte blanche* enjoyed by the oil companies up to now had to change; for example, the oil revenues received by the government were calculated from the companies' own figures. Moreover, as Angel J. Villas pointed out in a letter to Gómez the companies' profligate production practices were harmful to the country not only because a limited resource was being wasted, but also, more importantly at this juncture, because the country was losing revenue on this wasted oil since the companies paid tax only on the production figures submitted to the government, which were invariably their oil export figures. A department concerned solely with the oil industry needed to be created within the Ministry to inspect, control, and act as a watchdog over, *inter alia*, these production figures. Torres, however, picked an inauspicious time to put these policies into action for the difficulties facing the oil companies in the U.S. and the world-wide economic depression conspired to decrease Venezuelan production to the detriment of Treasury receipts.

Nevertheless, Luis Calvani, together with Jorge Aguerrevere

(technical adviser), drafted the oil regulations to the 1928 Oil Law[87] without, as had been the case in previous oil laws, the suggestions of the oil companies. The regulations had two main objectives: first, in order to avoid misunderstandings, to lay down clearly the legal requirements which concessionaires had to follow; and, second, to establish a more rigorous fiscal and technical inspection of the companies. In addition, the new regulations would also deal with the minor deficiencies of the law which experience had brought to light.

On 7 August 1930 the new oil regulations were enacted, bringing greater and stricter control of the oil industry. The central hub of this control would be the Inspectorías Técnicas de Hidrocarburos, which replaced the old Fiscalías de Hidrocarburos. The new Inspectorías would have the dual role of being 'technical and fiscal at the same time',[88] and would be staffed by qualified engineers and geologists well acquainted with oil legislation.[89] Many of the technical dispositions were to prevent competitive drilling, and were based on similar regulations controlling the American oil industry. Torres entrusted the organization of the new Inspectorías to Guillermo Zuloaga, a recently graduated doctoral student in geology from the Massachusetts Institute of Technology.

Three Inspectorías were established covering the country,[90] one at Maracaibo (headed by Guillermo Zuloaga and with five field inspectors), another at Coro (headed by Eneas Iturbe and with two field inspectors), and one at Maturín (headed by Pablo H. Carranza and with one field inspector).[91] The Inspectorías were strengthened when on 30 March 1931 the Oficina Técnica Central (headed by the Inspector Técnico General de Hidrocarburos) was created. According to Luis Herrera, the first Inspector-General, its function was to study in detail and advise the Minister on all matters relating to the oil industry.[92] Several of the inspectors were susceptible to receiving bribes from the oil companies. This was confirmed in 1932 by R. Alvárez de Lugo, who had been appointed by the government to look into these matters.[93]

The new regulations caused 'some perturbation among the oil companies to whom they seem to involve or admit a degree of Government interference which presupposes a numerically adequate staff of disinterested experienced officials such as this

country does not at present possess'.[94] The companies' opposition to the regulations was much greater than O'Reilly, the
British Minister, supposed, and from the very beginning there
was resistance against complying with the new regulations.
The companies' initial disagreement stemmed from the provisions concerning the elaboration of survey plans to be
presented to the Sala Técnica de Minas which the companies
considered contradictory to the 1928 Oil Law. Lago, for
instance, believed that the plans should just show with accuracy
the area controlled. Torres, however, thought differently,
insisting that the survey plans be presented as indicated in the
regulations. Leon C. Booker, the Manager of Exxon, did not
accept this request because the regulations could not stipulate
requirements different from those contained in the law. Thus
the fact that the plans were accepted under the law, and not
under the regulations, indicated to Booker that the regulations
did not comply with the law.[95] The point made by Booker,
Doyle of Shell and Smith of the Pantepec Oil Co. of Venezuela,
was that for the last two years the plans submitted by the
companies had been legally accepted. It was, therefore, impossible for the present regulations to make hitherto acceptable
plans unacceptable. In order to avoid any inconveniences,
Torres recommended Doyle and Booker to request the Ministry
to consider the plans as a special case outside the scope of the
regulations. This Smith had already done.[96] Torres would have
approved the plans as a conciliatory gesture, but the companies
refused to follow his suggestion. The companies were stalling for
time to permit them to draw up a Memorandum detailing their
objections to the regulations. Once again the companies formed
an action committee, this time composed of Doyle, Booker, and
E. Aguerrevere (of the Paraguaná Petroleum Corp.), which by-
passed Torres altogether, and presented their Memorandum
directly to President Pérez and Attorney-General Santos. Torres
remarked later that this was the first time the companies had
taken such drastic action 'without having sustained any damage
to their interests'[97] against the regulations which had been
legally enacted by the government of the country. Both government officials, however, referred the matter to him, and on 20
September he received a similar Memorandum from Doyle for
his perusal 'in which they detail their observations in a solemn

and exalted manner, signed by all the company managers as if it was the Treaty of Versailles'.[98]

The companies were confident that their views would prevail because, as O'Reilly reported, Gómez would not antagonize foreign interests on account of the worsening economic climate. In addition, it was known that Torres was at loggerheads with Ruben González, the Minister of the Interior, who was likely to lose no opportunity to discredit his colleague.[99] It was thus reasonable to assume that under these circumstances the 'judicious and united action of the companies will once more be successful in overcoming the difficulties with which they are threatened'.[100]

Torres, too, was confident of winning, and although he had expected some reaction from the companies, he was very much surprised at the extreme position adopted by them. In a letter to Zuloaga, enclosing a copy of the companies' Memorandum, he expressed his feelings thus:

I have clearly seen the disgust and unfounded alarm raised by the oilmen against the Regulations, and I even think that this is natural among people accustomed to guiding the government, as has been the case for many years. I do not claim that the Regulations are free of errors, nothing is perfect, and we lack the necessary technical skills, but from this to the alarm and fuss that has been raised there is an enormous gap: everything in the Regulations is procedural, leaving the core of the concessions intact. The time limit for a concession has not been shortened, taxes have not been increased, not a hectare reduced, nothing that affects deeply any concession has been done.[101]

The companies' main line of attack, as in the case of the buoy tax, was that the regulations altered the spirit of the 1928 Oil Law, and was therefore unconstitutional and not legally binding. Under the Constitution a set of regulations could not change or modify an existing law. In order to bring the regulations into harmony with the oil law, the companies suggested abrogating and amending the clauses they found objectionable as follows. Articles 18–46 imposed a series of new requirements with regard to mapping out the companies' concessions, which the companies were unwilling to accept because these would involve a great deal of alteration to the maps about to be submitted in order to convert their exploration concessions into exploitation concessions. They were, however, willing to provide detailed topographical maps in due course, but unwilling to alter the existing ones in the limited time available before the concessions expired, thus losing them by default. It was also apparent that if the regulations were strictly adhered to, then all maps submitted would be illegal as they would not have satisfied the new requirements. Similarly, if the maps were legal, then it was obvious that the regulations were illegal because they modified the existing law. Articles 62–6 established that companies needed to obtain permission from the government for the use *('establecer servidumbres')* of such items as water from Lake Maracaibo. Article 91 imposed the requirement of obtaining government permission to drill new wells on the companies' concessions, to which the companies objected strongly by virtue of their contracts. They were, however, prepared to notify the field inspectors when a new well was to be drilled, and to give them all the drilling details. Articles 80–9 imposed further restrictions on the companies' drilling programmes, establishing the minimum distance to be observed between wells. In addition, article 14 required the companies to build safety fire walls around oil storage tanks, and each tank was also to be separated by a minimum distance. The companies believed that this provision should not apply to tanks holding 'combustible oil', but only to tanks holding more than 3,000 tons of crude oil. Article 53 stated that before the Finance Minister could issue the tax return forms for the consumption tax on petroleum derivatives, the Minister of Development needed to receive a list compiled by the Inspector Técnico of all the oil products sold in the country, an impossible task since the consumption tax was levied at retail outlets. The companies requested that article 67 (which established that companies, when submitting their applications for equipment to be exonerated fom paying import duties, needed to specify the port of embarkation) be abolished. This was deemed impossible on the grounds that companies were not privy to the purchasing arrangements of their respective head offices. Article 87, which gave the inspectors access to the companies' tax returns, needed amending, and article 89, which restricted drilling rigs to a 100-metre minimum distance from the nearest house or workshop, needed modification as the companies could not control the overnight appearance of shacks near to their rigs. It would be more useful if the distances specified had been guidelines, and not rigid ordinances. Articles 98–101 dealt with the 'shutting-off' of gas, oil and water wells with cement, and the government demand that it be given five days' notice before the operation was concluded. The companies objected because it was felt that this was not viable and because delays would hamper their operations. The companies also wanted article 103 (prohibiting the wastage of gas by allowing it to escape into the atmosphere) abolished because they claimed they did not possess the technical expertise to conserve the gas, which was used to bring oil to the surface; and, moreover, there was no market for it in Venezuela. In addition, because of the world-wide depression, the companies found it unprofitable to invest in machinery which could extract natural petrol from the gas. The companies also wanted articles 105–6, which required the installation of a meter on each well to determine the amount of oil and gas produced to be abolished because, it was alleged, no adequate meter was available, and because it would only increase costs to the companies without any real benefits to either party. It was suggested that production could be inspected cheaply and easily by measuring the oil stored in the deposit tanks. For gas such a system presented difficulties, and it was therefore suggested that the old system by which the companies supplied the Ministry with their monthly production figures be retained. Finally, article 107, which called for an abandoned well to be plugged within a year, should be modified to allow the companies the discretion to carry out the operation when it was clear that the well was definitely non-productive.[102]

Although Torres was aware that the companies feared that the regulations were only the beginning of a government policy designed to achieve even greater control of the industry, he came to the conclusion, after carefully reading the Memorandum, that the companies' grievances were only 'silly claims which amount to the companies having to strictly comply with the requirements stipulated in the oil law'.[103] During the Cabinet meeting which was convened to discuss the companies' Memorandum, Torres, after a great deal of insistence, finally convinced his colleagues of his arguments by pointing out that the companies' real objection was to the government's right to acquire through its own 'officials the necessary information to protect its rights and those of individuals, and to follow the industry's progress, in order to obtain reliable data for fiscal purposes'.[104] The enactment of the regulations only created the adequate technical organization needed to administer and safeguard the oil law. He also explained that the companies' objections had ignored the close co-operation given by the government in stimulating the industry. Since 1918, when the first regulations relating to the oil industry had appeared, the laws concerning the industry had changed five times in order that they might keep abreast with the rapid development that had taken place and not be a burden to the industry.[105] Another privilege abused by the companies was the exoneration of import duties on machinery and equipment needed by the industry, a privilege only accorded by three other countries, viz, the U.S., France and Panama (the last two being non-oil producers), with Mexico only allowing this on a once-and-for-all basis (*Ley* 1901). In Venezuela this generous privilege had been blatantly misused with office equipment and other non-essential equipment being imported to the detriment of local importers, manufacturers and government revenues. Between 1920 and 1929 exonerations totalled Bs.233,359,462, while for the same period taxes paid amounted to Bs.171,952,126, a 'loss' of Bs.61,407,336, prompting Torres to remark that the government would have benefited more by levying import duties on the oil industry rather than by imposing taxes.[106] But it is worth remembering that the industry nevertheless significantly increased the volume of other imports which did pay customs duties. Consequently the country gained in net terms from the oil industry despite the high level of exonerations (see Table 25).[107]

Table 25   *Comparison between total oil income and tax exonerations granted to oil companies, 1924–35 and 1935–6 (in Bs.)*

	1924–35	1935–6
Oil income (including taxes)	540,508,666	71,558,430
Exonerations	361,708,749	25,875,662
Net oil income	178,799,917	45,682,768

Source: Pietri, *Lago . . . contra la Nación . . .* (Caracas, 1940), p. 324

In addition, companies such as Lago and Gulf Oil had been slow in reaching an agreement over the taxes owed to the government. The intention of the regulations was simply to establish a Departamento Técnico de Petróleo to look after the country's interests. It therefore appeared that the companies' intention was to 'promote unjustified alarm, instead of solving imaginary or feigned conflicts with the purpose of preventing the Nation from exercising its right, which it has taken with the regulations, to protect its national interests without in any way damaging those of the producers'.[108] Torres rejected the companies' claim that the regulations were unconstitutional because their enactment had not 'created, defined, nor interpreted rights; also further obligations had not been imposed, nor have those already established by law become more onerous'.[109] The objections raised to articles 18–46 were unsubstantial as from 1926 the standard topographical details on maps of the concessions, which previously included rivers, ponds, paths, houses, etc., had declined considerably. In many cases, when applying for exploitation concessions, the companies had not bothered to include maps, and when the Sala Técnica of the Ministry had requested them, the companies had refused to comply. There was, therefore, a need to specify the various features which needed to be included in the maps. The controversy raised by the companies over the completed maps since the enactment of the 1928 Oil Law was of their own doing because the Ministry had been willing to consider the maps.[110] In Torres' opinion it was necessary to justify the need for *servidumbres* before these were granted, and therefore articles 62–6 only established the conditions under which applications would be accepted, but it was understood that they would be provided free. Torres also felt that the nation, as owners of the oil before it reached the surface, was entitled to be informed about new drilling operations. Thus he did not see any harm in article 91, nor did he see that it contradicted article 22 of the oil law which recognized the right of concessionaires to exploit all minerals on their concessions, and to drill as many wells as they wished. Article 52 of the same oil law stated that oil companies had to exploit the oil 'following closely the scientific and practical principles applicable to the region',[111] but it did not oppose the concessionaires' exclusive right to extract oil or curtail their activities.[112] By the same token Torres also rejected the companies' objections to articles 80–9, and argued that because of the long-term detrimental effects of competitive drilling the government had the right to determine where companies drilled in their concessions. He pointed out that the companies in Maracaibo had made a similar agreement privately. He also rejected the companies request to amend article 14 because 'in Venezuela the majority of the fires in the oilfields started in the pick-up stations, that is the small storage tanks next to the wells, because being small they are unprotected, and are more liable to catch fire, and for it to spread'.[113] He found the objections raised to article 67 incredible[114] for he felt that the companies with close ties to their head offices could easily obtain the information required. The object of article 87 was not to obtain information from the company without the manager's consent or knowledge[115] as the law provided for the field inspector to arrive at the companies' offices during the first two days of each month. To ensure that the government received the proper tax returns from the companies, the field inspectors needed to gain access to the companies' accounts, especially when it was known that some managers had not been honest with the information previously supplied.[116] He considered the objections to article 89 unfounded since it was necessary to protect the lives and property of the people living near the oilfields. In many cases property had been lost in fires caused by oil spillages. It was also 'fantastic' that the companies were afraid of a shack going up overnight next to a rig because they could either purchase the property or, if it had been illegally erected, obtain a court order for its removal.[117] He also did not accept the companies' reasons for allowing natural gas to be flared, as the intention of the regulations was to encourage the use of gas in the extraction of crude oil, and the conversion of it into natural petrol. It was not exorbitantly expensive to install these conversion plants because several companies had previously invested in such equipment; for example, CPC (4 conversion plants installed), VOC (4), Lago (3), and British Controlled Oilfields Ltd (1). It was also nonsense to state that because natural gas did not have a ready market in Venezuela it should not be produced since the same argument could be applied to crude oil production. In both cases the principal market was the U.S. where Venezuelan crude competed successfully with others. Finally, Torres was adamant that it was the government which had to inspect production rather than the companies, and therefore rejected the objections to articles 105–6. Production tax was levied on oil extracted from the ground, and therefore the country's inspection was deficient because 'it was not carried out in the oilfields, at the wells themselves, where it is easy and effective'.[118] Contrary to the companies' claims, meters were readily available to do the job, being in general use in the U.S., Mexico, and Argentina. The amount of oil wasted through spillage, fires, and the general carelessness of the operators was large and went untaxed.[119] Thus, the nation's self-interest demanded a stricter and tighter control of the companies' production, which had to be measured at the well-head. Consequently, Torres ordered Zuloaga to make a study on how many field inspectors were needed to cover all producing wells without detriment to their other duties.[120] In addition, the companies were to maintain daily production figures which would be checked by the inspectors and which would be used for tax purposes in place of the export figures previously provided by the companies.[121] Finally, in order to prevent further loss of revenue because of the companies' loss of oil due to evaporation, fires, or leakages, it was established that the companies would settle their production tax during the middle of each month.[122]

After careful consideration the Cabinet approved Torres' fight with the companies, and on 18 October Torres formally replied to the companies, stating that their objections had been noted, but adding that only time 'and the sound aims that guide the government would be the best indicator of the future modifications needed to the regulations'.[123] At the same time he tried to allay their fears by reminding them that:

with regards to obtaining permission to drill a well, as the Ministry has already informed the Technical Inspectors, it is solely a way of the concessionaires notifying the government that it intends to exercise one of the rights granted by the concession; with reference to the survey plans, the offer is repeated that the ones already drawn up at the time the regulations were enacted will be approved.[124]

Despite this offer, the passive resistance, as Guillermo Zuloaga called it, of the companies continued. For instance, C.A. Velutini, on his arrival at Cabimas as Field Inspector of the oilfield, found that the companies 'were very displeased and it has taken a lot of hard work to obtain information from them'.[125] Each company reacted differently; for example, VOC's manager agreed to comply with the regulations with the exception of several clauses which were being studied in Caracas. The worst offender was Gulf Oil which directed Velutini to its Maracaibo office to obtain production figures. When, after consulting Torres, Velutini returned to check the company's books, he was informed that they did not keep any.

By the end of the year the companies reluctantly accepted the regulations and began to co-operate. But when Cayama Martínez replaced Torres in mid-1931, they again petitioned the Minister to amend the regulations. Cayama Martínez replied that this was impossible and that the regulations would continue to be applied in the same manner as in the previous eleven months. The government, however, while not amending the regulations, did not apply them rigorously in order to avoid offending the companies. E. Keeling, the Special British Envoy to Venezuela, reported in 1934 that the government was 'quite willing to leave aside the objectionable clauses without going as far as to withdraw them or to admit that they are in any way wrong'.[126]

ROYALTY FRAUD

Venezuelan oil was not traded in any market, the majority of the oil being transferred between subsidiaries of the same company. It was, therefore, difficult to determine the real value of oil in Venezuela. Up to 1929 the Venezuelan government had calculated this by deducting transport costs from prices realized in New York for crude oil of a similar grade to that produced in Venezuela. Torres faced yet another controversy with the companies arising out of the investigation conducted by the U.S. Tariff Commission in oil production costs in various parts of the world, which revealed that Gulf Oil and Lago had grossly inflated their transport costs between Venezuela and the U.S.,[127] thus paying smaller royalties because the tax was calculated on the net value of oil in Venezuela. The government reacted swiftly, and instructed Santos, the Attorney-General, to start proceedings against the oil companies, but his long illness which began at this time and ended in his death, prevented this from happening. It was decided that during the discussions to determine the real value of oil on which to levy tax, the companies would provisionally settle their taxes by paying Bs.1.50 per ton, the minimum established by law.[128] Despite this the companies did not respond to the government's pressing appeals for payment of taxes, and when Dr Villegas Pulido succeeded Santos as Attorney-General, the claim was taken up with renewed vigour.[129]

On 7 April 1931 Jordan H. Stabler of Gulf Oil requested further information on the problem. Two weeks later, on 20 April, the Minister replied that the company's settlement of its taxes was being reviewed. Four days later, on 24 April, Stabler outlined to Torres his company's proposals which he submitted formally the following day for the Minister's study. According to Stabler, when the company submitted its tax return slips the government could either accept or reject the tax base used to calculate its taxes. The Ministry, having accepted these calculations, then proceeded to hand over to the company the respective tax forms. This was the best proof available to demonstrate that the Ministry had accepted '*as irreversible,* the base proposed to pay the production tax'.[130] The government's willingness to accept a provisional settlement of Bs.1.50 per ton was interpreted as meaning that the company 'had to be *prepared to receive from the Government in the future . . . a counterproposal;* to determine the percentage basis on which to calculate the Government's production tax'.[131] It was strange, therefore, that a few months after the government had accepted the company's tax returns, it should want to revise the basis on which the government's tax share was calculated.[132]

Torres heard all these arguments with 'absolute impassiveness'[133] disagreeing that the Oficina de Liquidación did not have the right to revise its opinion; furthermore, he did not accept as definitive the proposal by the company to settle its tax using as its base figure the minimum established by law of Bs.1.50 per ton because no settlement had the character of a final payment. Tax payments were liable to re-examination by the Sala de Examén when there was doubt as to whether tax had been paid on the right basis. If this was not the case, 'every accounting system would go down the drain'[134] because there would be no redress for any erroneous decision made by the parties involved. The company's tax liabilities had been calculated wrongly because the company had supplied false information, with the result that 'all tax settlements referred to in your letter of the 24 of the current month can be revised'.[135] Torres submitted a Memorandum to the rest of the Cabinet for its decision on the 'legality and fairness of our claim'[136] against the companies.

The companies also took a firm stand on the issue. During a conversation on the U.S. Tariff Commission's findings and the companies' transport costs, Pannill, the head of Lago, informed Zuloaga that there had been no obligation on the part of his company to divulge such information, and that it was up to the government to discover it.[137] Gulf Oil did not accept Torres' views and ignored completely the central core of the Minister's argument, which was that the company had deliberately exaggerated its transport costs in order to pay less tax. Instead the company addressed its attention to a very legalistic interpretation of its liquidation of taxes. It was argued that Torres was ignoring the country's laws which stated that the job of the Oficina de Liquidación was to 'proceed to liquidate the tax in accordance with the declaration formulated by the contributor'.[138] In the specific case under dispute, the government and the company had together agreed on the percentage basis on which the production royalty would be settled. Once it had been agreed to by the two parties, the base rate then became immutable, and therefore could not be revised by the Sala de Examén, whose duty was limited (according to article 103(5) of the Ley Orgánica de la Hacienda Nacional) to checking 'the valuation and liquidation of taxes effected by the offices of the Administration'.[139] For this reason alone, the Minister could not make payment of taxes conditional on a future revision. Gulf Oil decided to call the government's bluff by stressing that if the Minister had sufficient cause for not accepting the company's tax returns, then these should be conveyed to the company by legal means. At the same time the company warned President Pérez of the dire consequences that the decision to revise its tax returns would have on the country's economy and on Venezuela's standing among international bankers and financial institutions.[140] Other interests closely associated with the oil companies began to campaign against Torres, and even sought his dismissal. Torres took this with equanimity, stressing to Arcaya that 'the intrigues of the oil companies do not concern me',[141] as his only wish was to serve Gómez and the country's interests to the best of his ability.

In order to obtain further information for the impending suit against the oil companies, Claudio Urrutía was sent to Washington to study the U.S. Tariff Commission's report in detail. For his part, Arcaya consulted the State Department to ascertain whether the U.S. government would support any claims arising from the dispute which the companies might bring against the Venezuelan government. The reply was that there was no 'danger of a claim by the American Government'.[142] In addition, Torres also pressed the companies to adhere to the regulations and to reorganize their accounts so that when the government requested transport costs or any other information 'the basis on which to liquidate taxes could be ascertained quickly and without discussion'.[143]

The matter was discussed at the last Cabinet meeting before Pérez's resignation as President. López Contreras, the War and Navy Minister, reflected the general consensus of Cabinet opinion in his belief that the only real solution to the problem was legal action against the companies. However, before this final drastic step was taken, every effort should be made to 'conciliate the Nation's interest in the intensive exploitation of its natural resources with the submission of the producers to the law'.[144] After receiving the support of his colleagues, Torres sent an extensive Memorandum to Lago and Gulf Oil to the effect that between February 1927 and January 1931 they had deducted from their tax schedules excessive transport costs between Venezuela and the U.S.; moreover, the 20 cents deducted for coastal shipping and storage from the price of oil was also considered unjustifiable because this was already included in the transport costs to the U.S. For these reasons the two companies owed the government Bs.61,452,899 in back taxes.[145] The government also wanted to revise its system of calculating royalty payments, as the one in use was too dependent on the figures provided by the oil companies, and it was almost impossible to ascertain the price charged for Venezuelan oil because it was not sold on the open market.[146]

The companies took little notice of Torres' request because they were waiting for the appointment of a new Cabinet. On 18 June, the same day that Congress unanimously elected Gómez to be President, they agreed to present their true transport costs to Torres. He was confident that he had won the case, and predicted that by the next week the 10-cents sales costs charged by the companies to the crude oil buyers would be dropped because 'they do not sell it, using it in their own refineries, and in the rare instances when they do sell it, I have proof that these costs vary between one and two cents per barrel, which is paid by the purchaser'.[147] This was Torres' last word on the subject as Minister of Development for, in the ensuing Cabinet re-shuffle, General Rafael Cayama Martínez was appointed in his place.

Lago seized upon this development. While not openly accepting that it owed the government anything, it suggested in early July a new way of calculating the base price of a barrel of crude oil. Taking the basic price to be $0.65 per barrel, then the difference in the U.S. between the actual price of the equivalent barrel and the average price set at $3.50 would be calculated. The same exercise would be repeated for Bunker C fuel oil (the average price used here would be $1.05 per barrel). These two figures would then be added to the basic price, and finally transport costs of $0.16359 from Zulia to Aruba would be deducted. The 7.5 per cent production royalty would be calculated on the resulting figure.[148] A committee, composed of Guillermo Zuloaga, Luis Gerónimo Pietri, and J.M. Betancourt Sucre, was appointed to study the proposal, which proved unacceptable to the committee and was subsequently rejected by the government.[149] The fundamental stumbling block still remained as the companies would not admit to outstanding debts for taxes. In September, at a Cabinet meeting, Cayama Martínez again presented the case for bringing legal action against the companies. Later Jordan Stabler notified Rafael Requena (Gómez's Secretary-General) that his company was now in agreement with the details contained in Torres' Memorandum on the matter, but that there were very powerful reasons, as Pannill had previously pointed out, for not complying with them.[150] On 25 September a new proposal for calculating the basic price on which taxes would be paid was presented by the company. After further discussions, the government determined that the company's tax could be liquidated under its methods but refused to accept the transport and sales costs as suggested because previous studies had demonstrated these figures to be highly inflated. Nevertheless, as the company's proposals would never produce revenue above the minimum required by law, it was decided that the company should pay instead the minimum established by the law.[151]

The government's decision was based on a study conducted by Gustavo Manrique Pacanins, Guillermo Zuloaga, E. Hermoso Domínguez, and E. Arroyo Lameda in April 1931 in which Venezuelan oil was considered to have the same commercial value as Bunker C fuel,[152] which was openly traded on the world oil markets. Consequently, to determine the commercial value of Venezuelan oil, it was only necessary to deduct transport costs ($0.37) from the price of Bunker C fuel ($0.90), giving it a value of $0.53 per barrel. A ton in Venezuela would therefore be priced at $3.3 or Bs.17.2, with the result that the government's 10 per cent royalty would be worth Bs.1.72 or Bs.0.28 below the minimum Bs.2 stipulated by law. In the case of Lago (which enjoyed a 25 per cent reduction in its production tax through its operations on Lake Maracaibo) the government still gained from the company's payments of the minimum required by law, as the 7.5 per cent royalty calculated on the figure given above would yield Bs.1.29 or Bs.0.21 below the minimum of Bs.1.50 established by law.[153]

Although agreement was reached on how the basic price should be calculated (which to all intents and purposes remained at Bs.1.50), the pressing point of taxes and arrears continued. For this reason, on 4 November 1931, Cayama Martínez sent the affidavit on tax evasion by Lago and Gulf Oil to the Attorney-General for him to prepare the country's case against the companies.[154]

OTHER PROBLEMS

Although tax evasion and the 1930 Oil Regulations continued to be the major issues facing the government at the time, other

minor questions came to light during the period. Gas production and its attendant difficulties also occupied Torres'
attention. The new regulations did force the companies to make
better use of the natural gas, but whether or not the companies
should pay tax on the natural gas produced was still unre-
solved. Torres commissioned Guillermo Zuloaga to make a
detailed study, and he concluded that gas which was burnt off
or used to extract natural petrol should be taxed.[155] Torres then
informed the companies that they had to pay tax on the gas
produced. Some companies such as the British Controlled
Oilfields Ltd tried to obtain a reduction because they pumped
their gas back into the wells.[156] Torres, however, rejected this
because the type of gas produced by the company was very
rich.[157] The Shell group of companies had to pay Bs.2 per
equivalent ton of gas produced which, 'by a wholly iniquitous
method, as illegal as it is harmful to the Nation's interests, the
companies have not paid in full because part of the total production was illegally deducted from the amount declared'.[158]
Other companies objected to paying the same tax on gas as on
crude oil but, after further discussions in December 1931, Lago
agreed to this, and later SOC (Venezuela) and Gulf Oil
followed suit.

Torres also took issue with the companies over the price of
petrol. There had been complaints that the price was too high
in a country which was the second largest oil producer in the
world. The minimum price of petrol observed in the country
was Bs.0.45 per litre in Barquisimeto, but a check in 23 other
cities revealed much higher prices, reaching Bs.1 per litre in
Encontrados.[159] In Argentina, which only managed one-
twentieth of Venezuela's crude oil production, the price of
petrol was Bs.0.40 per litre, and even in such non-oil producing
countries as the U.K., France, Italy and Germany the price was
cheaper than in Venezuela. In the U.S. petrol produced using
Venezuela crude sold at Bs.0.20 per litre after a sales tax of
Bs.0.10.[160] It was known that the cost of petrol production in
Venezuela was Bs.0.06–0.08 per litre, which with a sales tax of
little over Bs.0.03 meant that consumers in the country were
being overcharged. Torres commissioned the Raymond
Concrete Pile Co. to make an independent cost analysis for a
12,000 barrel per day refinery, similar to the one operated by

CPC at San Lorenzo. The results showed that it cost Bs.1.1352 to refine one barrel of crude oil, which produced 20 litres of petrol, 20 litres of gas oil, and 100 litres of fuel oil. The cost per litre of petrol was by then Bs.0.0676, but the other derivatives which were a by-product, sold at Bs.11.7 per 100 litres in Caracas, and in Maracaibo at Bs.6.3 per 100 litres, more than adequately covering the total costs of taxing, extracting and refining the oil of Bs.3.2516.[161] Appalled at this state of affairs, Torres in December 1930 complained to the companies about the high cost of petrol and requested them to reduce it forthwith.[162] Although the companies argued at first that, on account of the country's small market and transport difficulties (with high transport costs), they were unable to sell petrol cheaply, they did agree after a while to reduce it by Bs.0.05 per litre.[163]

INTEREST IN A NATIONAL REFINERY GROWS

Interest in establishing a national refinery which would produce cheap petrol intensified. Already in 1928 Aquiles Iturbe had proposed to purchase the government's royalty oil at $0.04 per barrel higher than the price offered by the companies. Iturbe planned to establish a large refinery at Turiamo to produce cheap petrol and diesel.[164] In November 1929 P.L. Manning proposed a similar project, guaranteeing the government the same prices paid by the companies for the royalty oil but with the added bonus of a 10 per cent profit share in the refinery.[165] Both projects, however, were rejected, but in 1930 a close political aide of Gómez arrived in Caracas with one similar to those outlined above. Gómez was willing to accept the project because it offered $0.05 above the companies' own price for the royalty oil. According to Calvani, this could be done because they knew that Lago and Gulf Oil were deducting grossly inflated transport costs from the price of oil, thus reducing the government's royalty tax.[166] Torres and Calvani did not agree with the project, and the former was even willing to tender his resignation if the government accepted the proposal. He did, however, feel that the government could obtain better prices for its royalty oil if it was taken in kind and then sold to the highest bidders.[167] In addition, long-term contracts could be entered into with Brazil, Uruguay, Argentina, Spain, France, and Italy which imported large quantities of oil. Torres also enquired of Rincones, the Venezuelan Consul-General in New York, the cost of a 2,000 barrel per day refinery and the prices obtained on the open market for oil similar to that of Venezuela.[168] Calvani convinced Gómez that the proposition was not a good one, and persuaded him to support Torres' view of selling the royalty oil to the highest bidder. As a first step towards achieving this end, large deposit tanks would be constructed at Turiamo where the royalty oil would be stored, and from where it could be easily distributed. Once oil was at Turiamo, many proposals for a refinery would follow, and the government could accept the one which offered the best conditions. Calvani, however, underestimated the interest in such a project at Turiamo, for the number of schemes put forward during the early 1930s increased dramatically.

One of these was that submitted by P.S. Luigi, who for twelve years had worked as Assistant Manager and later as Manager in various refineries in Europe. His proposal was the erection of a 10,000 barrel per day refinery which would employ 6,000 people,[169] and which would purchase the government's oil at a price 20 per cent above the market price of oil. The government would also receive an annual rent of Bs.66,616,000 and own 51 per cent of the 75,000 ordinary shares of Bs.1,000 issued by the company.[170] The rest would be subscribed by local interests, and in order to allow small investors a chance to subscribe to the company's equity, arrangements would be made with a local bank to allow them to acquire one Bs.1,000 share in twelve monthly instalments of Bs.83.33.[171] Luigi would also establish storage tanks at Turiamo and transport oil from Maraciabo to the port at his own expense. The government, however, rejected the scheme. Subsequently, Luigi re-submitted the scheme, again unsuccessfully, on 6 November 1930.[172] Similar proposals were submitted without success by the Venezuelan Royalties Corp., Southwestern Engineering Corp., Otto Scott Estrella (a Venezuelan with New York bankers backing his proposal), Oscar Schnell, Harry W. Schumacher, Charles Freeman, Oscar Irizarry Calder (Petroleum Refractioning Corp.), and José R. Osuma Lucena.

Torres explained the government's reluctance to enter into an agreement with a private commercial venture to exploit its royalty oil when he rejected Freeman's project. These objections were based on legal and economic arguments. Articles 44 and 45 of the Ley Orgánica de la Hacienda Nacional prohibited the sale of the government's royalty oil,[173] and the government's solvency would be virtually at the mercy of the company which controlled the royalty oil since the government would be handing over control of its second largest revenue earner without any guarantee of obtaining a similar return in the future. Under the present system the government was at liberty to opt for payment in kind or cash, 'a system which should be maintained as the last defence of its oil reserves'.[174] The schemes put forward would create complications in revenue collecting by establishing another intermediary. Guillermo Zuloaga felt that it was necessary to put an end to the notion of a central refinery using royalty oil because government revenues would suffer at the expense of the proponents of the scheme.[175]

The difference in price between Venezuelan oil and other sources of oil at New York gave Zuloaga the notion that Venezuela should adopt the Romanian system of auctioning its state oil to the highest bidder every three months, with the proviso that the producing company had first refusal on the oil provided it paid a price equal to the highest bid. But it was also recognized that owing to the world recession and to the low demand for oil the operating companies in Venezuela would join together to place the country 'between the devil and the deep blue sea'[176] by not bidding above the prevailing prices. There still existed, however, the possibility of selling the government's oil on the European markets of Spain, France or Germany. Zuloaga made a preliminary study of the French market alone, and found that the traditional sources of supply for the independent refiners would fall short by 750,000 tons, which could be adequately covered by Venezuela's royalty oil supplies, which in 1929 amounted to 780,039 tons. The advantage of this system was the increase in government revenues and the greater commercial and political independence in oil matters given to the country.

In spite of the disadvantages, the attraction for the government of establishing a refinery at Turiamo to process its royalty oil still remained. The government was not new to such a venture as from 1914 onwards it had successfully administered and exploited the country's coal deposits. The government, however, did not have a marketing monopoly as British coal held the largest share of the market.[177] The government, therefore, commissioned the Raymond Concrete Pile Co. to make a preliminary study of Turiamo and to advise on the cost of building a refinery there, later calculated at $5.5 million.[178] Rincones, the Venezuelan Consul-General in New York, also made a suggestion to Torres concerning their possible partnership in a refining venture using a new invention called Wade's Hydrogenation Process which produced oil very inexpensively. Torres declined the offer.[179]

Although the government did not build a refinery at Turiamo, private entrepreneurs still persisted in trying to obtain the contract. On 2 July 1931 L.C. Chase, supported by the U.S. Steel Co., submitted his industrial scheme for developing the Turiamo port which called for an oil refinery to be supplied with royalty oil.[180] In partnership with Hoyles Jones of Tulsa, Oklahoma, Preston McGoodwin sought to enter into an agreement with the government to refine, transport, and market royalty oil.[181] The company would purchase the oil at Bs.18.20 per ton, and would hand over 13 per cent of its net profits to the government.[182] At the same time Colonel Robert W. Stewart acquired Schumacher's scheme for a refinery at Turiamo. Stewart's company, to be called C.A. Nacional de Industrias Petroleras (CANIP), would refine and market royalty oil over a period of 40 years, and would also hold the domestic oil market monopoly. The company would have an authorized capital of Bs.40 million ($7 million), of which 55 per cent would be subscribed by the government and the remaining 45 per cent by Stewart. The Board of Directors would be made up of six members, of which three would be nominated by the government and three by Stewart, with the president of the company (Stewart) holding the casting vote. The company would later branch out into other aspects of the oil industry, and into banking, agriculture, transportation and construction.[183] The proposed contract was handed to Pedro R. Tinoco, the Minister of the Interior, for his opinion. Tinoco thought the contract not only too onerous but also illegal, since the operating companies had the contractual right to market the oil produced in the country. Thus if a law was enacted preventing this, the companies could claim breach of contract, and a number of large claims for compensation would soon follow.[184] Another of his objections rested on the fact that Stewart would receive nearly half the profits from the government's royalty. Under the present arrangement the government was guaranteed 10 per cent of the oil's value, but it was very doubtful whether the value added to the oil by CANIP would fully compensate for the loss of profits going to Stewart. CANIP would also face stiff competition both in the domestic and world markets, reducing profits to a minimum; moreover, in the domestic market, a price war could follow, benefiting the consumers but harming the government's oil revenues. A further drawback was that the company's supply came exclusively from the operating companies in the country, meaning that its ultimate success rested on the continual growth of oil production. If the government was convinced that it was not receiving its full 10 per cent of the oil produced in the country, it could institute a strict check on the companies to ensure that it did. Under CANIP's contract, the government's revenues depended on CANIP's profitability. In view of the impending law suits to be brought against Lago and Gulf Oil, Tinoco also felt that the scheme could be interpreted as a reaction by the government against the companies. Finally, he concluded that if the government wanted to refine its own oil in its own refinery, then it should engage 'American experts as employees, and not as partners'.[185] Tinoco's report sealed the fate of Stewart's project, which was rejected by Gómez on account of the insistence on establishing a domestic oil monopoly which was unconstitutional.[186]

THE GOVERNMENT RETREATS

The government by 1932 appeared set for a major confrontation with the oil companies. It was apparent that it sought new ways of increasing revenue, and even contemplated the establishment of a refinery which would compete directly with the oil companies. It was also about to start legal proceedings against two of the three largest oil companies in the country, while it appeared that Schumacher, who had re-activated the 1917 Hodge claim against CPC, would be successful, thus depriving Shell of one of its major subsidiaries in the country. The prospects for the established oil companies did not seem good, but the government would falter due to important external factors affecting the industry. It was well aware that the worldwide economic depression had led to a reduction in oil production and revenues. In addition, in the U.S., oil production continued to rise, and in 1931, in order to forestall the imposition of a U.S. import ban on foreign oil, the companies operating in Venezuela reduced production by 25 per cent. Arcaya calculated that this would reduce Venezuela's oil revenues from Bs.63 million to Bs.48 million.[187] This agreement was only a temporary measure for that year. As this was occurring, the House of Representatives approved congressman Garber's resolution to extend the coverage of the Hawley-Smoot Act to allow the Tariff Commission to investigate the cost difference between native and foreign oil production.[188] It found in 1932 that the average cost of U.S. domestic petroleum delivered to the Atlantic seaboard during 1927–30 was $1.90 per barrel, while the average cost of Maracaibo crude delivered there was $0.87 per barrel, and for all other foreign oil (including Venezuelan oil) it was $1.15 per barrel.[189] For this reason independent oil producers in the U.S. pressed for a tariff on imported oil of $0.42 to $1.00 per barrel.

The Venezuelan government was dismayed at these events as the imposition of a tariff would entail a severe cutback of the country's production. E.A. van Cleck was employed by the Venezuelan authorities to lobby Congress to stop the imposition of the tariff,[190] while Pedro Arcaya was busy rallying support for Venezuela at the State Department. Nevertheless, on 6 June 1932 the Senate finally approved tariffs on imported oil of $0.21 per barrel of crude oil and miscellaneous derivatives; $1.05 per barrel of petrol and other motor fuels; $1.68 per barrel of lubricating oil; and $0.01 per pound of paraffin and other petroleum wax products. The U.S. Treasury, however, decided that 'crude oil in bond, to be used in making refined products for exports, and fuel oil to be used as supplies for ships engaged in foreign trade could be imported free of tax',[191] which meant that bonded foreign oil could be refined in the U.S. and re-exported. This allowed Gulf Oil and Exxon to supply a large part of their eastern seaboard markets with Venezuelan oil. In addition, Venezuelan oil increased its share of the European market by replacing part of the U.S. domestic oil exported previously to Europe (see Table 26). Venezuela's share of Britain's oil market also increased dramatically during the 1930s from 17 per cent at the beginning of the decade to 49 per cent in 1939.[192] In the U.S., too, after 1932, Venezuelan crude oil still entered in large quantities, although petrol imports were negligible after the imposition of the tariff. At first, though, foreign imports declined, due almost entirely to the economic depression, from 88 million barrels in 1931 to 44 million barrels in 1933,[193] but then gradually increased, averaging, for 1937–8, 59 million barrels.[194]

Table 26   *Source of crude oil imports into Europe, 1928–33*
(in per cent)

Year	U.S.	Venezuela and Dutch West Indies	Romania	Russia	Persia
1928	38.8	13.4	7.8	8.4	14.2
1929	34.8	12.5	8.5	10.3	13.7
1930	33.5	13.0	10.3	12.3	13.6
1931	27.3	14.5	12.5	14.4	13.4
1932	21.5	15.3	13.4	15.8	13.6
1933	18.6	21.4	13.6	11.1	13.0

Source: adapted from 'Shifts in European supply and the Iraq Oil', *Petroleum Press Service*, 1:5 (1 August 1934), 1–3 (p. 2)

The development of the Venezuelan oil industry, and the experience of the 1932 U.S. tariff on foreign oil revealed to the Venezuelan government its vulnerability and lack of real control of the industry, and its dependence on the continual growth of production, which in turn depended on the requirements of the oil market of the industrialized U.S. and Europe. Venezuela's oil revenues, then, depended on the country's competitiveness and attractiveness to the oil companies inducing them to continue producing oil, for under the contractual obligations entered into by both parties, the companies could retain their valuable concessions while only producing the bare minimum of oil, much to the country's financial detriment. The effect of the U.S. tariff on foreign oil was to make Venezuela more vulnerable because the competitive edge which the industry had over the U.S. was narrowed. The feeling among government officials was that the oil companies could leave as quickly as they had come, or considerably reduce production, especially during a long dispute, since the country did not hold the same attractive prospects as before. Oil revenues, which accounted for 30 per cent of the government's total revenues, would, therefore, fall, with disastrous political and economic effects for the Gómez dictatorship. Consequently, the Venezuelan government's demands of, and power over, the oil companies lessened, and it was forced to lay aside its suit for tax evasion against Lago and Gulf Oil. Moreover, on 15 June 1932, the Federal Court of Cassation dismissed Schumacher's claim against CPC,[195] and on 6 October 1932 the same court decided that Carvallo's concession, which claimed part of certain river beds and streams running through VOC's and CPC's concession, was valid only up to midstream.[196]

THE LAST THREE YEARS

During the last three years of Gómez's rule no fundamental changes took place in the industry or in government policy. A new oil law was enacted on 1 July 1935 which had as its sole innovation article 44, which allowed the government to deal directly with the oil companies in matters relating to oil concessions. The government was, however, always looking at ways to increase its oil revenues, and when Rufino González Miranda suggested that he could raise oil revenues by Bs.10 million by applying more rigorous accounting procedures at the Ministry of Development, Gómez quickly took up the suggestion, commissioning him to make a fuller and more detailed study.[197] González reiterated the clause in the 1930 Oil Regulations concerning the inspection of producing wells by Ministry inspectors, but preferred a daily inspection since tax was paid on oil reaching the surface and any part which evaporated or was spilt represented a loss to the Exchequer. In order to achieve maximum revenue from the government's production royalty the various oil grades needed to be taken into account because the 'sweeter' high grade oils fetched higher prices. By fixing a single price for its royalty oil the country was not receiving the full value of its production royalty.[198] It was, therefore, of great importance to determine the real value of Venezuela's oil to the companies. Another area where savings could be made was in the strict application of the companies' duty free imports. González realized that the companies had abused this privilege, and that the large operating companies had benefited the most (see Table 27). He adduced that between 1926 and 1933 the companies' exonerations from customs duties amounted to Bs.255,558,615, whereas oil revenues during the same period amounted to Bs.304,214,231, a difference in favour of the government of only Bs.48,655,616.

Table 27   *Comparison of certain companies' tax and exoneration schedules, 1926–33 (in million Bs.)*

Company	Taxes	Exoneration	Difference
CPC	58.7	40.2	18.5
VOC	73.9	64.2	9.7
CDC	10.5	8.8	1.7
SOC (Venezuela)	21.4	11.3	10.1
Lago	56.7	53.4	3.3
Gulf Oil	41.4	47.9	− 6.5
British Controlled Oilfields Ltd	4.4	6.8	− 2.4
TOTAL	267.0	232.6	34.4

Source: AHMSGPRCP 20–31 Julio 1934, Rufino González Miranda, 'Memorándum', undated

The seven companies listed in Table 27 contributed 88 per cent of total oil revenues and received 91 per cent of the industry's exoneration from customs duties. In some cases such as Gulf Oil and the British Control Oilfields Ltd, the government was a net loser, and in the case of the Shell group the gains were relatively small because they paid low tax rates. Nevertheless,

González felt that savings could be made by curtailing the amount of imports which illegally entered the country, such as spare parts for cars, electric fittings, construction fittings, wood, etc., which in 1933 amounted to Bs.1.4 million. Further savings could be obtained by stopping items being imported by companies which were not directly used in oil production, such as timber and cement for house construction, amounting in 1933 to Bs.4.1 million. If closer scrutiny had been carried out on imports, the government in 1933 would have gained Bs.5.4 million (44.9 per cent of the total exonerations for the year in question), or 14.7 per cent of the total oil revenues. These savings would be relatively greater during the time when companies were not expanding their operations, and consequently not importing drilling equipment. To carry out his recommendations González suggested that a Central Accounting Office be organized in the Ministry for the purpose of keeping separate accounts for each company. Instead of lumping together oil revenues under one general heading *(Renta Minera)*, as was the case, the new accounts would show the total tax paid by each company, and the distribution of this among the various tax categories. The exercise would also be adopted for customs exonerations so that each company would have a record of what items enjoyed this privilege. The importance of this procedure, according to González, was that the government could determine at any time the net benefit accruing to the country from each company.[199]

González's plans did not meet with universal approval. Herrera, head of the Inspectoría General de Hidrocarburos, raised several important objections. According to him, the Servicio de Inspectores already checked the companies' production by determining the amount of oil kept in storage tanks; daily well inspections would be too expensive, costing the country Bs.3.6 million annually in wages paid to the additional inspectors employed,[200] and to those who would have to be employed to check the storage tanks. But this was all unnecessary because the inspectors had access to the companies' own production books.[201] The companies could also provide the government with all the details needed on the density of oil produced, but in many cases (Shell for example) the grade of oil produced was immaterial because a fixed tax on production was paid. It was also impossible to determine the real value of Venezuelan oil because it was not 'quoted in any market'.[202] Herrera stressed that during the present world economic depression Venezuela received more for its oil by charging the minimum stipulated by the law than it would by calculating the government's 10 per cent royalty using world oil prices.[203] Herrera, however, did not object to keeping more detailed accounts of the oil industry, and González's recommendation in this area was taken up by the Ministry of Development.

In the ensuing years, the government maintained its vigilance on the living and working conditions of the oil workers. In 1933, for example, Cayama Martínez requested SOC (Venezuela), Gulf Oil, Lago and the Tocuyo Oilfields of Venezuela Ltd to provide their work force with adequate medical assistance.[204] C.A. Velutini, the Field Inspector in Falcón, also drew attention to SOC (Venezuela)'s lack of accommodation for its workers at its Cumarebo oilfields.[205] The following year the Minister reported that the companies were acting in a more socially responsible manner because the number of accidents had decreased and the sanitary conditions of the companies' accommodation, and the health welfare of the workers, had been improved considerably.[206]

The activities of the oil companies also began to increase during 1933, surpassing all estimates in 1934, and achieving record oil production in 1935. The amount of construction in the oilfields also expanded, with SOC(Venezuela) opening a new pipeline and terminal at Guiría. The additional activity also increased accidents, such as an enormous half-metre thick oil spillage near La Concepción in Zulia. While not accepting responsibility for the accident, Lago donated Bs.50,000 for the victims.[207] The country also appeared to be on the verge of a new oil boom. Since 1930 Gómez had limited the number of concessions granted, and in 1935 Arcaya counselled that the government should continue to resist the temptation of selling concessions at the first price offered. It was best to let competition among the oil companies push prices up,[208] and with decreasing oil supplies in the U.S. the companies would invariably turn to Venezuelan concessions, 'competing among themselves to acquire them, thus increasing the price'.[209] Arcaya, therefore, advised Gómez to 'stop granting concessions to individuals who will later transfer them to the companies',[210] because 'the nation should endeavour to establish competition among the companies by awarding concessions directly to them, and thereby increasing the price paid for concessions'.[211] Higher prices would be achieved the longer the delay in awarding new concessions. Nevertheless, at the end of 1935 the government granted SOC(Venezuela) 286 exploration concessions in Anzoátegui and Monagas, for which the company agreed to pay much higher tax rates than before, the initial exploitation tax being set at Bs.25–30 per hectare, and the government's royalty at 12–15 per cent.[212]

The oil industry was thrown into temporary chaos after Gómez's death, and there were reports that the new government of López Contreras was adopting an 'anti-foreign policy . . . mainly against the oil companies with the idea of squeezing them into making good the extravagance of the Government's finances'.[213] There was a feeling among local oil managers that the country could develop into a second Mexico, and there was open talk of pulling out altogether. However, the enactment of the new oil law in September 1936 allayed their fears, and the companies appeared to be satisfied as no representations to the diplomatic corps were made, and there were 'indications that the American companies are starting the same somewhat mad scramble for concessions in Venezuela as they did in Mexico some 10 or 15 years ago'.[214] Apart from the large concessions taken up by SOC(Venezuela), Socony-Vacuum also took up 300,000–500,000 hectares in eastern Venezuela, paying $2 million.[215]

As the industry consolidated and production increased, the government became more aware of the industry's complexity and the various means, both legal and illegal, open to the companies to minimize their taxes and their obligations. Reaction to this was swift in that the government shifted its objectives away from the need to attract foreign capital to ensuring that the companies complied with their contracts. The need for legislation to safeguard a more vigorous supervision of the industry in all its phases was promptly fulfilled. But for a country which under Castro's dictatorship (when Gómez was Vice-President) had suffered the skulduggery of the New York and Bermúdez Co., and the French Cable Co., and which also had a great deal of mining experience, there was an almost naive belief in the power of the law to prevent companies from acting in a dishonest and fraudulent manner. The repeated requests by the Ministry of Development for the companies to provide better health and living conditions for their native work-force, and the fortuitous discovery of the royalty fraud perpetrated by Lago and Gulf Oil disproved this belief. A necessary prerequisite for the effective control and supervision of the industry was the creation of an efficient oil bureaucracy that would deal on equal terms with the industry. Such a bureaucracy was envisaged in the 1930 Oil Regulations, but the government, to its own peril, failed to support this venture adequately, and must be severely criticized for this. In other respects it was slightly more successful: for example, it achieved a pyrrhic victory when two companies agreed to pay Bs.12 million in buoy taxes owed, obtaining in return a Bs.1 per ton reduction in tax for 25 years; a modest decrease in petrol prices was achieved; and the government was only restrained from taking two other companies to court for bilking its royalty tax payments by the worsening world economic situation.

At the same time the companies also reacted promptly to face the government not as heretofore as single concerns, but together under the strong leadership of the large operating companies, winning their first significant victory in 1927 when Buckley's Salinas Bay project was cancelled.[216] This co-operation, which from now on would be one of the significant features of the industry, was a strong force when the companies objected to the 1930 Oil Regulations, and put forward their recommendations for amending them. However, despite the inauspicious time chosen, the government too, was more confident of its own position and, in sharp contrast to the policy adopted eight years earlier, resisted the companies' pressure, with the result that the regulations stood. It must be noted, however, that although they were not rigorously applied, the threat of their enforcement was always present, and they formed an effective framework for supervision of the industry. The experience of the last ten years of Gómez's rule defined to both parties concerned the limits of their own influence. The government was constrained by not having an effective oil bureaucracy, and by external market forces on which it had no influence. The companies' action in Venezuela also depended on what occured in their major U.S. and European markets, but they also recognized the importance of maintaining good relations between government and industry. But this principle, however, was only extended to areas in which the government held adequate control. There is no doubt that the companies held the upper hand in the relationship for they were better organized (both technically and managerially); in most cases they could count on diplomatic assistance from powerful countries, and, most important of all, Gómez needed the oil revenues more than the oil companies (especially during the Depression years) needed the oil. In all the disputes between the companies (with the exception of the companies which held ambiguous contracts awarded during Castro's time or during Gómez's first presidency) there was never any question of forfeiting their concessions as a penalty for not complying with the country's laws. Although the government looked for ways of increasing its revenues (at one point it contemplated seriously refining and marketing its own royalty) it was content to let the companies develop the industry while it acted as a quasi-watchdog and reaped the oil revenues.

Conclusion

On the assumption of power on 19 December 1908, Gómez presented the country with an economic plan to rehabilitate it after his predecessor's chaos. An important element in this plan rested on a healthy and striving mining industry, which would spearhead economic development and provide the government with an independent source of income. Thus, from the very beginning the importance of this sector of the economy was recognized. It is therefore no surprise to find that keen interest was aroused when several British oil companies, after lengthy negotiations, entered the country. The government was well aware of the importance and value of oil, and was determined to get a fair return from the companies, as well as to establish effective control over the industry. We find that recommendations of this nature were made very early during Gómez's regime by people closely associated in the Cabinet with the oil industry, and that unsuccessful attempts were made prior to 1919 to increase oil production. Despite his personal links with the industry, the possibility that it would provide Gómez with an independent source of income made him aware that in order to ensure his own political survival he would need to get as much out of the companies as possible in taxes. Thus from 1914 onwards the government pursued a consistent policy of securing a larger tax return from the companies. But Gómez's interest did not end here, as he was also interested in protecting ordinary people from the excesses of the oil companies. These two requirements meant that government control and supervision of the industry would have to be increased. It is doubtful, however, whether Gómez or anybody else in 1922 foresaw the phenomenal growth that would take place and make the country in the course of six years the second largest oil producer

213

in the world. The swiftness of this increase caught the government, both regionally and nationally, unawares, and the strained public administration was unable to cope. A bureaucracy that could effectively manage the oil industry on equal terms was envisaged in the 1930 Oil Regulations, but the government failed to support the venture, and although the economic recession of the early 1930s is partly to blame because it demonstrated poignantly the constraints acting on the government, nevertheless the government can be severely criticized for its failure to create a force that would control and supervise the industry. It is thus possible to conclude that Gómez and his government were well aware of the importance of oil, and took a keen interest in its development, long before the industry had established itself in a significant way in the country; that Gómez sought by all possible means to increase the return from the industry; that despite close personal and family ties with the industry, out of political expediency he did not become an instrument of the oil companies; that Gómez established an effective framework to control and supervise the industry; and, that considerable protection was extended to workers in the industry.

Appendix

1 *Distribution of concessions acquired by companies, 1908–36*

Parent holding company	Acquisition of concessions						Total concessions held under company			Number of concessionaires	
	D	Tr	F	Tr	T	Tr	Gross	Tr	Net	Gross	Net
Alamo Oil Corp.	3	3	2	1	14	10	19	12	5	16	5
Algeo Oil Concessions Ltd	19	19	33	26	16	7	68	52	16	33	15
Amerada Corp.	0	0	16	0	6	0	22	0	22	15	15
American Controlled Oilfields Ltd	0	0	9	3	0	0	9	3	6	3	3
American Maracaibo Co.	0	0	10	2	6	0	16	2	14	12	11
Andes Petroleum Corp.	4	0	93	0	2	0	99	0	99	24	24
Atlantic Refining Co.	18	0	24	0	4	0	46	0	46	20	20
British Controlled Oilfields Ltd	123	75	84	0	1	0	208	75	133	19	19
Brokaw, Dixon and Gardner	0	0	19	0	1	0	20	0	20	20	20
Caribbean Consolidated Oil	2	0	3	0	0	0	5	0	5	3	3
Caribbean Oilfields of Venezuela	0	0	1	0	0	0	1	0	1	1	1
Central Mining and Investment Corp. Ltd	0	0	16	9	0	0	16	9	7	1	1
Chacín and Lample	2	1	2	0	0	0	4	1	3	2	2
Cities Service Co.	59	0	6	0	1	0	66	0	66	5	5
Compagnie des Asphaltes de France	0	0	7	7	0	0	7	7	0	1	0

1 (*cont.*)

Parent holding company	Acquisition of concessions						Total concessions held under company			Number of concessionaires	
	D	Tr	F	Tr	T	Tr	Gross	Tr	Net	Gross	Net
CVP	304	216	24	17	1	1	329	234	95	24	5
Consolidated Oil Corp. (Sinclair)	62	0	393	0	42	0	497	0	497	82	82
Dakota Oil and Transport Co.	0	0	18	0	72	0	90	0	90	15	15
Dubuc and Ochoa	0	0	1	0	0	0	1	0	1	1	1
Falcón Petroleum Corp.	0	0	2	0	2	0	4	0	4	2	2
A.H. Garner	0	0	0	0	6	4	6	4	2	3	1
General Asphalt Co.	0	0	75	0	0	0	75	0	75	3	3
Gulf Oil	2	0	143	0	100	0	245	0	245	73	73
H. Cheney Hart	0	0	12	1	4	0	16	1	15	15	15
C.A. Hidrocarburos Mocacai	0	0	1	0	0	0	1	0	1	1	1
Edwin B. Hopkins	0	0	20	4	0	0	20	4	16	6	6
Intercontinent Oil Concessions	0	0	39	8	18	0	57	0	57	11	11
Lagomar Oil Concessions	0	0	27	8	0	0	27	8	19	1	1
Maracaibo Oil Exploration Co.	0	0	31	0	0	0	31	0	31	7	7
Yervant Maxudian	0	0	0	0	1	0	1	0	1	1	1
Alfred Meyer	37	19	309	100	102	6	448	125	323	104	84
George F. Naphen	0	0	0	0	3	0	3	0	3	1	1
National Venezuelan Oil Corp.	16	0	3	0	0	0	19	0	19	4	4
New England Oil Corp.	0	0	6	0	4	3	10	3	7	8	5
Omnium Oil Development Co. Ltd	2	0	6	0	0	0	8	0	8	3	3
Oriental Oil of Venezuela	0	0	29	29	0	0	29	29	0	1	0

Orinoco Oilfields Ltd	0	0	72	71	0	0	72	71	1	2	1
Pantepec Oil Co. of Venezuela	49	37	122	36	43	21	214	94	120	78	44
C.A. Petróleo Altagracia	0	0	1	0	0	0	1	0	1	1	1
C.A. Petróleo de Cantaura	0	0	5	0	0	0	5	0	5	1	1
C.A. Petróleo de Paraguaná	0	0	1	0	1	0	1	0	1	1	1
C.A. Petróleo del Lago	0	0	0	0	5	0	5	0	1	1	1
C.A. Petróleo Las Mercedes	0	0	0	0	0	0	1	0	5	1	1
C.A. Petróleo Minerales Rio Pauji	1	0	0	0	0	0	7	0	1	0	0
Pure Oil Co.	7	0	0	0	0	0	3	0	7	0	0
Royal Dutch-Shell Group	0	0	3	0	3	0	23	0	3	3	3
Seaboard Oil Co.	0	0	20	1	0	0	1	1	22	17	16
Simms Petroleum Co.	1	1	0	0	0	0	1	1	0	0	0
Société Française de Recherches au Venezuela	0	0	0	0	1	0	1	0	1	1	1
Socony-Vacuum	0	0	48	0	25	0	73	0	73	13	13
The South American Co.	0	0	0	0	7	7	7	7	0	1	0
SOC(California)	11	11	16	0	37	4	64	15	49	27	25
Exxon	613	0	471	0	288	0	1,372	0	1,372	187	187
Sun Oil Co.	3	0	62	6	1	0	66	6	60	29	28
The Texas Corp.	0	0	8	0	8	0	16	0	16	9	9
Trans-America Petroleum Ltd	0	0	0	0	2	0	2	0	2	1	1
Ultramar Exploration Co.	0	0	12	11	15	10	27	21	6	7	3
Union Oil Co. of California	7	0	0	0	25	6	25	6	19	14	10
United Venezuelan Oil Corp.	0	0	0	0	0	0	7	0	7	0	0
Unity Petroleum Co.	0	0	2	1	0	0	2	1	1	1	1
Val de Travers Asphalt Paving Co. Ltd	0	0	0	0	7	0	7	0	7	1	1
Venezuelan Consolidated Oilfields	6	3	5	0	6	0	17	3	14	2	2
Venezuelan Development Co.	0	0	1	1	0	0	1	0	1	1	0
Venezuelan Eastern Oilfields Ltd	0	0	36	24	0	0	36	24	12	6	2

1 (*cont.*)

Parent holding company	Acquisition of concessions						Total concessions held under company			Number of concessionaires	
	D	Tr	F	Tr	T	Tr	Gross	Tr	Net	Gross	Net
Venezuelan Fuel Oil Syndicate Ltd	0	0	1	1	0	0	1	1	1	0	0
Venezuelan Holding Co.	3	0	1	1	0	0	4	1	3	3	0
Venezuelan Investment Co.	0	0	0	0	4	4	4	0	4	4	3
Venezuelan Northern Oil Concessions	0	0	2	0	0	0	2	0	2	2	1
Venezuelan Oil Exploration Co.	0	0	0	0	1	1	1	0	1	1	1
Venezuelan Syndicate Ltd	0	0	13	5	5	5	18	10	8	8	4
Vimax Oil Co.	0	0	0	0	12	0	12	0	3	12	3
TOTAL	1,354	385	2,366	365	902	84	4,622	834	3,788	964	829

Key: D concessions acquired directly from the government
      F concessions acquired from original title-holders
      T concessions acquired from third-party intermediaries
      Tr transfers

Source: calculated from ADCOTHMEM, 'Historial de concesiones de Hidrocarburos', Vols. 1–34, Files 1–16,620

2 Distribution of concessions acquired by Venezuelan intermediaries, 1908–36

Name of intermediary	Acquisition of concessions						Total concessions held			Number of concessionaires under intermediary	
	D	Tr	F	Tr	T	Tr	Gross	Tr	Net	Gross	Net
Alamo, C.J.	0	0	7	0	3	0	7	0	7	1	1
-Ambard, F.R.	11	10	1	1	0	0	12	11	1	1	0
Angulo, J.G.	6	0	4	4	0	0	10	4	6	4	0
-Andrade, I.	13	5	3	0	0	0	16	5	16	1	1
Aranda, H.	9	0	1	0	0	0	10	1	5	1	1
-Arcaya, C.	0	0	2	1	0	0	2	10	1	2	1
Arreaza Alfaro, P.	6	6	0	0	4	4	10	1	0	1	0
Barberi, L.E.	0	0	1	1	0	0	1	1	0	1	0
Bello Torres, M.	0	0	7	4	0	0	7	4	3	1	1
Benedetti, L.	0	0	7	6	0	0	7	6	1	2	1
-Brandt, A.	0	0	2	2	0	0	2	2	0	2	0
-Bueno, A.	0	0	3	3	2	1	5	4	1	3	1
-Calcaño Vethancourt, O.	6	0	5	0	0	0	11	0	11	1	2
Capriles, A.	12	6	8	0	0	0	20	6	14	2	2
Capriles, J.M.	11	5	3	3	0	0	14	8	6	3	0
Capriles, M.J.	24	23	27	10	1	0	52	33	19	17	13
Capriles, R.	17	16	0	0	14	0	31	16	15	1	1
-Carvallo, L.	7	6	4	4	0	0	11	10	1	1	0
-Casanova, J.V.	0	0	0	0	2	2	2	2	0	1	0
Chacín, P.A.	12	8	4	3	0	0	16	11	5	4	1

2 (*cont.*)

Name of intermediary	Acquisition of concessions						Total concessions held			Number of concessionaires under intermediary	
	D	Tr	F	Tr	T	Tr	Gross	Tr	Net	Gross	Net
Chapman, J.	0	0	2	0	0	0	2	0	2	2	2
Chellini, M.A.	0	0	2	0	0	0	2	0	2	1	1
-Colmenares Pacheco, J.A.	16	0	34	1	0	0	50	1	49	7	6
Corao, M.	6	0	12	12	0	0	18	12	6	2	0
Corrales, M.	0	0	8	8	0	0	8	8	0	1	0
-De Capriles, A.	0	0	7	7	3	1	10	8	2	5	2
-De Faria, A. Arcaya	1	0	1	0	0	0	2	0	2	1	1
-De Fleites, M.	0	0	1	0	0	0	1	0	1	1	1
-De Gómez, J. Revenga	0	0	7	0	4	0	11	0	11	3	3
-De Iturbe, C. Ceiba	0	0	1	1	0	0	1	1	0	1	0
-De Lanz, C. Vallenilla	0	0	2	0	0	0	2	0	2	1	1
-De Maldonado, L. Bello	0	0	1	0	0	0	1	0	1	1	1
-De Noel, I.C.	0	0	1	0	0	0	1	0	1	1	1
-De Parra, V.M.	0	0	1	0	0	0	1	0	1	1	1
De Sanoja, A. Jaén	0	0	2	2	0	0	2	2	0	1	1
-De Urrutia, C.V.	0	0	1	0	0	0	1	0	1	1	1
Ferris, J.	13	0	15	7	1	1	29	8	21	16	6
Fonseca, D.	0	0	2	0	0	0	2	0	2	1	1
-Gómez, J.V.	0	0	1	0	0	0	1	0	1	1	1
-González Rincones, E.	0	0	4	0	0	0	4	0	4	2	2

Gorrín, J.	0	0	13	13	0	13	13	0	1	0
–Guerrero, F.A.	0	0	5	5	0	5	5	0	5	0
Gutiérrez Alfaro, P.	0	2	4	0	0	4	0	4	1	0
Heny, C.	2	0	23	10	0	25	12	13	4	3
Herrera, G.	0	0	42	19	0	42	19	23	7	4
Hurtado Rondón, M.	0	1	0	0	0	1	0	1	1	1
Insausti, J.I.	1	1	1	0	1	1	0	1	1	1
Iturbe, G.	4	4	0	0	0	2	2	1	0	0
Kunhardt jun., H.R.	0	0	12	0	0	16	4	0	9	9
Lehamn, P.S.	0	0	0	0	0	1	0	12	1	1
León, R.	0	0	0	0	0	1	0	1	1	1
León, T.E.	0	0	5	5	0	5	5	0	1	1
López, Castro, R.	0	0	1	0	1	1	0	1	20	7
–López Rodríguez, J.	0	0	18	11	0	19	12	7	1	0
Madriz, M.	0	0	2	2	0	2	2	0	1	0
Manrique Pacanins, G.	0	0	5	5	0	5	5	0	1	1
Márquez, J.M.	0	0	1	0	1	1	0	1	1	0
Márquez, M.A.	0	0	0	0	0	7	1	0	1	0
Martín, F.I.	0	0	7	7	1	4	7	0	2	2
–Maury, C.H.	0	0	3	0	1	2	1	3	4	3
Mendoza, C.L.	0	0	1	1	0	2	1	1	1	1
Mendoza, F.	0	0	2	2	0	2	2	0	2	0
Mendoza Borges, R.	0	0	2	2	0	13	2	0	1	0
–Mendoza Fleury, L.A.	0	0	12	12	0	2	12	0	13	1
Mercado, L.	1	1	1	1	0	1	1	1	1	1
Mondolfi, U.	0	0	1	0	0	1	1	0	2	0
Moros, J.D. and Colmenares, M.	0	0	0	0	0	1	0	1	1	1
Negretti, C.	0	0	1	0	0	1	0	0	1	1
–Noel, A.	0	0	0	0	0	1	0	1	1	1
–Olavarría Matos, J.A.	0	0	7	7	0	7	7	0	2	0

2 (cont.)

Name of intermediary	Acquisition of concessions						Total concessions held			Number of concessionaires under intermediary	
	D	Tr	F	Tr	T	Tr	Gross	Tr	Net	Gross	Net
Otero Vizcarrondo, H.	8	7	1	0	0	1	10	8	2	2	1
Pacanins, A.	0	0	0	0	0	0	1	0	1	1	1
Perea, J.A.	24	11	15	6	0	0	39	17	22	4	2
Perea, S.	6	6	6	6	6	6	18	18	0	2	0
Pérez, J.D.	9	7	4	0	0	0	13	7	6	1	1
–Pirella Páez, A.	52	28	0	0	1	0	53	28	25	1	1
Pocaterra, J.	0	0	9	4	0	0	9	4	5	2	2
Prosperi, P.	0	0	1	0	0	0	1	0	1	1	1
Ramos Baglais, M.	5	0	4	4	0	0	9	4	5	1	0
Rojas, F.	0	0	1	0	0	0	1	0	1	1	1
Rodríguez Lange, J.H.	0	0	1	1	0	0	1	1	0	1	1
Rodríguez, J.S.	0	0	1	1	0	0	1	1	0	1	1
Ron Pedrigue, M.L.	0	0	1	1	0	0	1	1	0	1	0
Sanabria, G.	0	0	1	0	0	0	1	0	1	1	1
–Sira, L.	6	6	24	24	0	0	30	30	0	4	0
–Sira, M.	0	0	24	24	0	0	24	24	0	4	0
–Tagliaferro, L.	1	0	12	0	0	0	13	0	13	2	2
de Tovar, J.M.	0	0	16	3	0	0	16	3	13	10	7
–Urrutía, V., C.	4	2	6	5	0	0	10	7	3	2	1
–Urrutía, C.	8	2	3	3	0	0	11	5	6	2	0

–Urrutía, L.	5	0	12	12	0	0	17	12	5	1	0
–Urrutía, R.	6	6	6	6	0	0	12	12	0	1	0
Valdivieso Montano, A.	0	0	1	1	0	0	1	1	0	1	0
Velazco Castro, D.	0	0	6	0	0	0	6	0	6	1	1
Zuloaga, N. and Nevett, G.	0	0	7	7	0	0	7	7	0	4	0
Zuloaga, N.	0	0	2	2	0	0	2	2	0	2	0
TOTAL	312	168	529	291	49	20	890	479	411	244	117

Key: – *gomecista* intermediaries
D concessions acquired directly from the government
F concessions acquired from original title-holders
T concessions acquired from third-party intermediaries
Tr transfers

Source: calculated from ADCOTHMEM, 'Historial de concesiones de Hidrocarburos', Vols. 1–34, Files 1–16,620

3 *Distribution of concessions acquired by foreign oil intermediaries, 1908–36*

Name of intermediary	Acquisition of concessions						Total concessions held			Number of concessionaires under intermediary	
	D	Tr	F	Tr	T	Tr	Gross	Tr	Net	Gross	Net
Berg, N.	0	0	2	2	0	0	2	1	1	1	1
Brown, W.F.	0	0	6	6	0	0	6	6	0	3	0
Hart, R.C.	0	0	6	0	0	0	6	0	6	1	1
Flipper, H.O.	0	0	5	5	1	1	6	6	0	2	0
Fontanier, A.F.	0	0	1	1	0	0	1	1	0	1	0
Freeman, C.	0	0	1	1	0	0	1	1	0	1	0
Herman, C.E.	0	0	5	5	0	0	5	5	0	3	0
Howard, O.R.	0	0	15	14	0	0	15	14	1	2	1
Hutchings, C.A.	4	0	0	0	4	0	8	0	8	1	1
Knight, F.H.	0	0	1	0	0	0	1	0	1	1	1
Muller, E.	0	0	0	0	1	1	1	1	0	1	0
Ortiz, H.G.	0	0	1	0	0	0	1	0	1	1	1
Porter, E.F.	0	0	12	12	54	54	66	66	0	11	0
Rathbone, C.H.	0	0	3	3	3	0	3	3	0	1	0
Robb, E.E.	0	0	0	0	1	0	1	0	1	1	1
Schumacher, H.W.	0	0	6	3	0	0	6	3	3	2	1
Shea, H.H.	0	0	0	0	3	3	3	3	0	1	0
Smith, W.W.	0	0	18	18	3	0	21	18	3	8	2
Stewart. H.R.	0	0	5	5	0	0	5	5	0	1	0

Stewart, J.W.	0	0	3	3	0	0	3	3	0	1	0
Wallace, W.W.	0	0	0	0	1	0	1	0	1	1	1
Wiltsee, E.	0	0	3	1	0	0	3	1	2	3	2
TOTAL	4	0	93	78	68	59	165	137	28	48	13

Key: D concessions acquired directly from the government
    F concessions acquired from original title-holders
    T concessions acquired from third-party intermediaries
    Tr transfers

Source: calculated from ADCOTHMEM, 'Historial de concesiones de Hidrocarburos', Vols. 1–34, Files 1–16,620

Notes

Introduction

1 U.S. House of Representatives, 'Production costs of crude petroleum and of refined products', *House Document No. 195*, 72 Cong. 1 Sess., 1932.
2 E.H. Davenport and S.R. Cooke, *The oil trusts and Anglo-American relations* (London, Macmillan and Co., 1923).
3 Calculated from: Venezuela, Ministry of Mines and Hydrocarbons, *Venezuelan petroleum industry. Statistical data* (Caracas, Ministry of Mines and Hydrocarbons, 1966), p. 1.
4 U.S. Senate, 'Cost of crude petroleum in 1931', *Senate Document No. 267*, 71 Cong. 3 Sess., 1931.

1 The dawning of an era

1 Venezuela (comp. Luis Correa), *El General Juan Vicente Gómez. Documentos para la historia de su Gobierno* (Caracas, Lit. del Comercio, 1925), 'Mensaje que el General Juan Vicente Gómez, Presidente Provisional de la República, presenta al Congreso Nacional en 1910', 19.4.10, p. 55.
2 Cf. José Antonio Linares, *El General Juan Vicente Gómez y las obras públicas en Venezuela, 19 de diciembre de 1908–4 de agosto de 1913* (Caracas, Lit. y Tip. del Comercio, 1916).
3 Venezuela, *Gómez. Documentos*, 'Mensaje que el General Juan Vicente Gómez, Presidente de la República, presenta al Congreso', 29.5.09, p. 24.
4 Ibid., 'Mensaje del General Juan Vicente Gómez para el Congreso en sesiones extraordinarias', 15.10.11., p. 4.
5 MinFo, *Memoria 1907–8*, Exposición, p. vi.
6 AHMSGPRCP 24–31 Dic. 1908, José R. Colina to J.V. Gómez, 26.12.08.
7 *DDCS*, Mes 2(16), 21.6.10.
8 Ibid.
9 Vice-Consul Guy Gilliat-Smith, 'Diplomatic and Consular Reports. Venezuela: Report for the year, 1909–10 on the trade of Venezuela and the Consular District of Caracas', *PP* XCVII (1911) 767–807, (p. 773).
10 FO 199/208, Sir V. Corbett to Sir E. Grey, 27.11.10.
11 AHMSGPRCP 1–14 Mayo 1909, Ascanio Negrette to Gómez, 12.5.09.
12 Cf. *RLDV*, Vol. 36, Docs. 11,413 and 11,414, p. 208, and 11,440, pp. 236–7.
13 MinFo, *Memoria 1912*, Exposición, p. viii.
14 AHMSGPRCP 1–14 Mayo 1909, C. Contreras to Gómez, 11.5.09.

227

15 MinFo, *Memoria 1914,* Exposición, pp. xii–xiii.
16 AHMSGPRCP 1–14 Mayo 1910, J. Proctor to Gómez, 11.5.10.
17 FO 199/251 Letter, unsigned, to J.A. Tregelles, 16.11.09.
18 PA, A.C. Veatch, 'Columbia–Venezuela Project – general plan of campaign', 24.7.12.
19 Walter R. Skinner, *The Oil and Petroleum Manual* (London, 1912), pp. 99 and 143.
20 MinFo, *Memoria 1911,* Exposición, p. vii.
21 PA, Ribon, 'Memorandum', 2.10.12.
22 Ibid.
23 FO 368/755, F.D. Hartford to Grey, 8.1.12.
24 Bs.200,000 for each concession transferred to the Bermúdez Co., and to the Caribbean Petroleum Co. Ltd (AGTU, Rafael Max. Valladares to Proctor, 4.3.29.).
25 AGTU, Valladares to Proctor, 4.3.29.
26 Ibid., at the end of the letter Proctor ratifies what is stated therein, dated Madrid, 6.6.29.
27 FO 368/755, Hartford to Grey, 8.1.12.
28 Sir Henri Deterding, *An international oilman* (London, Ivor Nicholson and Watson Ltd, 1934), p. 97.
29 Ralph Arnold, George A. MacGready and Thomas W. Barrington, *The first big oil hunt: Venezuela 1911–1936* (New York, Vantage Press, 1960), p. 85.
30 Interview: Antonio Aranguren Fonseca, 14.9.77.
31 MinFo, *Memoria 1913,* Exposición, pp. vii–viii.
32 ADCOTHMEM, 'Historial de Concesiones de Hidrocarburos', Files 14,010–6.
33 Venezuela, Consejo de Gobierno, *Memoria, 1914,* A. Lotowsky *et al.,* 'Informe', 24.3.14. pp. 295–7, (p. 294).
34 Ibid., pp. 296–7.
35 AMFHC 1932 *(sic),* A.J. van Oosteveen to Pedro Emilio Coll, 3.4.14.
36 MinFo, *Memoria 1914,* Exposición, p. xii.
37 AHMSGPRCPU, Santiago Fontiveros to Gómez, 11.6.15.
38 MinFo, *Memoria 1914,* Exposición, p. x.
39 MinFo, *Memoria 1915,* Exposición, p. x.
40 Ibid., p. x. A similar agreement was reached with the Bermúdez Co.
41 AHMSGPRCP 16–31 Enero 1916, Horacio Castro to Gómez, 24.1.16.
42 Ibid.
43 Ibid.
44 Ibid.
45 Ibid.
46 AHMSGPRCP 15–30 Abril 1916, Castro to Gómez, 28.4.16.
47 Ibid.
48 MinHa, *Memoria 1917,* Doc. 43, Julio César Silva to Ministro de Hacienda, 16.2.17, p. 47; and Doc. 44, E.W. Hodge to Ministro de Hacienda, 16.2.17, pp. 47–8.
49 MinHa, *Memoria 1917,* Doc. 46, p. 50.
50 AHMV 1917–20, 'Informe sobre la situación jurídica de The Caribbean

Petroleum Company', 24.2.17.
51 AHMSGPRCP 1–28 Feb 1917, Alejandro Urbaneja to Ministro de Fomento, 27.2.17.
52 AHMSGPRCP 1–15 Marzo 1917, V. Márquez Bustillos to Gómez, 17.3.17.
53 MinFo, *Memoria 1917*, Vol. 1, Exposición; and MinHa, *Memoria 1917*, Docs. 48 and 50, pp. 51–2. For further discussion see B.S. McBeth, *Royal Dutch-Shell vs. Venezuela* (Oxford Microfilm Publications, 1982), pp. 90–9.
54 *DDCD*, Mes 1(1), 19.4.14.
55 *DDCD*, Mes 1(2), 23.4.15.
56 Other members of the same committee included Raúl Capriles, assistant to Dr Ezequiel Vivas, Gómez's Secretary-General, whose family in later years would be influential in oil circles; and Andrés Mata, who together with A.J. Vigas founded the *El Universal* newspaper. In 1917 Rafael Requena was appointed to the Senate Permanent Development Committee.
57 Article 25 (1909, 1910 Mining Law), art. 29 (1918 Mining Law), art. 28 (1920 Oil Law), art. 7 (1921 Oil Law), and art 5 (1922, 1925, 1928, 1935 Oil Law).
58 AHMSGPRCP 11–19 Oct. 1934, Felipe Alvárez Cienfuegos to Gómez, 17.7.34.
59 Ibid.
60 Ibid.
61 AHMSGPRCP 1–15 Marzo 1917, Pedro Guzman to Gómez, 10.3.17.
62 Despite this, Juancho's influence would be felt throughout the case which dragged on until 1924 (cf. McBeth, *Royal Dutch-Shell vs. Venezuela*, pp. 182–215).
63 Ibid., pp. 174–81.
64 DS 831.6363/C23/7, P. McGoodwin to Secretary of State, 29.10.18.
65 AHMSGPRCP 16–31 Enero 1917, C.J. Rojas to Gómez, 18.1.17.
66 Ibid.
67 Juancho later acquired oil concessions in his own right over his estates in Miranda State (ADCOTHMEM, 'Historial de Concesiones de Hidrocarburos', Files 9,994–7).
68 AHMSGPRCP 15–30 Marzo 1916, Fontiveros to Gómez, March 1916.
69 AGTC 1917, José Vicente Gómez to G. Torres, 24.1.17.
70 MinFo, *Memoria 1916*, pp. 99–100.
71 AHMSGPRCP 15–31 Julio 1917, unsigned, to Márquez Bustillos, 31.7.17.
72 'Una poderosa industria nacional, la gasolina y el kerosene de Venezuela', *El Nuevo Diario* (Caracas), 15.9.17.

2 The legal framework

1 MinFo, *Memoria 1916*, Exposición, p. ix.
2 Ibid., pp. xvii–xviii.
3 Ibid., p. xix.
4 Ibid., p. xx.
5 Cf. B.S. McBeth, 'Juan Vicente Gómez and the oil companies' (D.Phil. Diss., Oxford University, 1980), ff. 89–94.
6 AHMCOP 177, Márquez Bustillos to Gómez, 13.11.17.

7 Ibid.
8 MinFo, *Memoria 1917*, Vol. 1, Exposición, p. xvi.
9 Ibid., p. xviii
10 Ibid., pp. xviii–xix.
11 Ibid., p. xix.
12 Ibid., p. xix.
13 Ibid., p. vii.
14 AHMCOP 177, Márquez Bustillos to Gómez, 22.3.18.
15 AGTCOP 9 and AHMSGPRCP 16–31 Mayo 1922, Torres to Gómez, 10.5.22.
16 MinFo, *Memoria 1919*, Exposición, p. xii.
17 Ibid., pp. xii–xiii.
18 MinFo, *Memoria 1918*, art. 5(2) of the law.
19 Ibid., art. 5(3) of the law.
20 Ibid., Exposición, p. vii.
21 *RLDV*, Vol. 41, Doc. 12,635, p. 23.
22 MinFo, *Memoria 1918*, Torres to Rafael Max. Valladares, 7.2.18, Doc. 134, p. 200.
23 AHMV 1917–20, Lewis J. Proctor, 'Memorándum que presenta The Caribbean Petroleum Company sobre haber cumplido la obligación de explotar sus concesiones y sobre la negativa de los Guardaminas a expedirle las planillas para el pago de impuestos y certificados de comienzo de explotación', 12.4.18; and MinFo, *Memoria 1918*, Doc. 140, pp. 209–19 (p. 210).
24 MinFo, *Memoria 1918*, Doc. 140, pp. 209–19 (p. 210).
25 AMF, Proctor to Torres, 21.5.19.
26 AMF, Proctor to Torres, 12.6.19.
27 AMF, Proctor to Torres, 21.5.19.
28 AMF, C.W. May to Torres, 20.5.19.
29 AMF, Ch. Couchet to Torres, 16.5.19.
30 These were: Bernabé Planas, Vigas, Aranguren, and Jiménez Arraiz.
31 Proctor, 'Memorándum', p. 211.
32 Ibid., p. 214.
33 Ibid., p. 214.
34 Ibid., p. 215.
35 Ibid., p. 215.
36 Ibid., p. 215.
37 Ibid., p. 217.
38 FO 371/4622, C. Dormer to Lord Curzon, 16.3.20.
39 AHMSGPRCP 1–30 Julio 1920, Torres, 'Memorándum', 10.7.20.
40 Ibid.
41 These rights were that Vigas would automatically subscribe, at no further cost to him, to 25 per cent of any increase in the company's equity.
42 AHMSGPRCP 1–30 Abril 1919, Pedro César Domínici to Gómez, 15.4.19.
43 AGTAPM 1919, to Torres, unsigned and undated.
44 AHMSGPRCP 1–30 Junio 1919, Torres to Gómez, 3.6.19.
45 AGTCOP 5, Torres to Rafael Hidalgo Hernández, 12.12.19.

46 AGTCOP 5, Torres to Vicente Lecuna, 18.12.19.
47 Pedro Manuel Arcaya, *Memorias del Doctor Pedro Manuel Arcaya* (Madrid, Talleres del Instituto Geográfico y Catastral, 1963), pp. 161–2.
48 Ibid., p. 165.
49 AGTAPM 1919, Arcaya, 'Informe sobre el régimen del petróleo', 31.12.19.
50 Ibid.
51 Ibid.
52 Ibid.
53 Arcaya, *Memorias*, p. 166.
54 Cf. Douglas H. Carlisle, 'The organization for the conduct of foreign relations in Venezuela, 1909–1935' (Ph.D. Diss., University of North Carolina at Chapel Hill, 1951), ff. 213–14; and 'Esteban Gil Borges, Canciller (1919–1920)', *BAHM*, 15:76 (Julio–Dic. 1973), 147–226 (pp. 174–5).
55 MinFo, *Memoria 1921,* 'Nómina de Contratos Vigentes, Petróleo y Carbón', pp. 104–13.
56 AHMSGPRCP 1–31 Julio 1919, Julio F. Méndez to Gómez, 17.7.19.
57 MinFo, *Memoria 1919;* and ADCOTHMEM, 'Historial de Concesiones de Hidrocarburos', Files 12,998–13,001 and 13,003–4.
58 Walter R. Skinner, *The Oil and Petroleum Manual* (London, 1920), 'Maracaibo Oil Exploration Corp.', p. 105.
59 AHMSGPRCP 16–22 Dic. 1924, A.H. McKay to Gómez, 17.12.24.
60 AHMSGPRCP 1–30 Marzo 1920, Torres to Gómez, 17.3.20.
61 AHMSGPRCP 1–30 Marzo 1920, Torres to Gómez (José V.) 17.3.20.
62 MinFo, *Memoria 1921,* Chas. R. Eches, 'Informes que la British Equatorial Oil Co. Ltd. presenta', 15.1.22.
63 AGTC Abril–Julio 1920, Santos M. Gómez to Torres, 23.4.20.
64 AGTC Enero–Marzo 1920, Juan C. Gómez to Torres, 24.1.20.
65 AGTC Enero–Marzo 1920, Elías Rodríguez to Torres, 15.1.20.
66 AHMSGPRCS 1–31 Ago 1919, B.G. Maldonado to E. Urdaneta Maya, 1.9.19.
67 AGTAPM 1919, Arcaya, 'Informe'.
68 FO 199/275, W. Seeds to Sir A. Chamberlain, 2.1.26; and U.S. Senate, *American petroleum interests in foreign countries,* Hearings before a Special Committee Investigating Petroleum Resources, 79 Cong. 1 Sess., 1946, p. 264.
69 'Terrenos petrolíferos de Venezuela y Colombia', *BCCC* 9:81 (15 Ago. 1920), 805–6.
70 Vicente Lecuna, *El historiador Vicente Lecuna y nuestra riqueza petrolera* (Caracas, Fundación Eugenio Mendoza, 1975), p. 7.
71 AHMSGPRCS Mayo 1920, F. Monsalve Terán, 'Comentarios al Discurso de W.G. McAdoo sobre los créditos de la América Latina', undated.
72 AGTC Enero–Marzo 1920, Gómez to Torres, 7.2.20.
73 MinFo, *Memoria 1920,* Exposición, pp. xv–xvi.
74 Lecuna, *Historiador,* p. 7.
75 Ibid.
76 DS 831.6363/25, McGoodwin to Secretary of State, 5.4.20.

77 Ibid.
78 Ibid.
79 Lecuna, *Historiador*, p. 7.
80 AGTCOP 6, Torres to Gómez, 20.5.20.
81 Ibid.
82 Lecuna, *Historiador*, p. 10.
83 DS 831.6363/70, McGoodwin to Secretary of State, 23.7.21.
84 AHMSGPRCP 16–31 Ago. 1920, Arcaya to Gómez, 17.8.20.
85 MinFo, *Cuenta 1921*, p. 4.
86 MinFo, *Memoria 1921*, pp. 50–76.
87 AHMSGPRCP 1–30 Julio 1920, Torres, 'Memorandum', 10.7.20.
88 DS 831.6363/41, McGoodwin to Secretary of State, 9.7.20.
89 DS 831.6363/475, *Statement of facts with respect to the Vigas Concession, its transfer to the Colon Development Company Limited, and the interests of the Carib Syndicate Limited therein* (New York, The Evening Post Job Printing Office Inc., Feb. 1929), Annex 11, pp. 113–16.
90 AHMSGPRCP 1–30 Julio 1920, Torres, 'Memorándum', 10.7.20.
91 FO 371/4622, Dormer to Curzon, 16.3.20.
92 MinFo, *Memoria 1919*, Vol. 3, Doc. 1, p. 327.
93 AHMSGPRCP 1–30 Julio 1920, Torres, 'Memorándum', 10.7.20.
94 AGTCOP 5, Torres to Gómez, 25.3.20.
95 FO 371/4622, Dormer to Curzon, 16.3.20.
96 DS 831.6363/Teleg., McGoodwin to Secretary of State, 7.4.20.
97 Edwin Lieuwen, *Petroleum in Venezuela* (Berkeley, University of California Press, 1954), p. 21.
98 FO 371/4622, CDC to Head Office, 28.4.20.
99 FO 371/4622, CDC to Head Office, 11.5.20.
100 FO 371/4623, J.C. Clarke to Foreign Office, 12.5.20.
101 AHMSGPRCP 1–30 Mayo 1920, Domínici (Pedro C.) to Gómez, 12.5.20.
102 FO 371/4623, Clarke to Foreign Office, 12.5.20. Gómez received this telegram since Domínici, on 7 June, acknowledged receiving this telegram from him, stating: 'I await correspondence; I have dealt with the matter'. (AHMSGPRCP 1–14 Junio 1920, Domínici (Pedro C.) to Gómez, 7.6.20.).
103 AHMSGPRCP 1–30 Mayo 1920, Domínici (Pedro C.) to Gómez, 12.5.20.
104 AHMSGPRCP 15–30 Junio 1920 *(sic)*, Duncan Elliott Alves to Gómez, 17.5.20.
105 Ibid.
106 DS 831.6363/29, C.K. MacFadden to Bainbridge Colby, 18.5.20.
107 DS 831.6363/27, Alvey A. Adee to MacFadden, 29.5.20.
108 FO 371/4623, Foreign Office to Dormer, 18.5.20.
109 FO 371/4623, Dormer to Foreign Office, 17.6.20.
110 FO 371/4623, Foreign Office to Dormer, 17.6.20.
111 AHMSGPRCP 1–30 Julio 1920, Dormer to Foreign Minister, 10.6.20.
112 AHMSGPRCP 1–30 Julio 1920, Torres to Gómez, 14.7.20.
113 DS 831.6363/41, McGoodwin to Secretary of State, 9.7.20.
114 AHMSGPRCP 1–30 Julio 1920 *(sic)*, Dormer to Foreign Minister, 10.6.20.

115 Ibid.
116 Ibid.
117 DS 831.6363/33, McGoodwin to Secretary of State, 11.6.20.
118 Lieuwen, *Petroleum,* p. 22.
119 DS 831.6363/26, McGoodwin to Secretary of State, 11.6.20.
120 DS 831.6363/26, McGoodwin to Secretary of State, 26.4.20; and DS
    831.6363/58, McGoodwin to Secretary of State, 5.5.21.
121 DS 831.6363/26, McGoodwin to Secretary of State, 26.4.20.
122 FO 371/4623, Dormer to Curzon, 23.7.20.
123 DS 831.6363/93, Murray to Welles, 27.4.21.
124 DS 831.6363/33, F.M. Dearing to McGoodwin, 6.5.21.
125 DS 831.6363/33, McGoodwin to Secretary of State, 11.6.20.
126 DS 831.6363/58, McGoodwin to Secretary of State, 4.5.21.
127 DS 831.6363/33, Dearing to McGoodwin, 6.5.21.
128 DS 831.6363/33, DS to U.S. Legation, 24.6.20.
129 FO 371/4623, Dormer to Curzon, 4.6.20.
130 AHMSGPRCP Feb. 1923 *(sic)*, Juan J. Mendoza to Torres, 10.8.20; and
    Ibid., P. Itriago Chacín, 'Consulta', 23.7.20.
131 AGTCOP 6, Torres to Gómez, 11.8.20.
132 FO 371/4623, Dormer to Curzon, 9.10.20.
133 Ibid.
134 FO 371/4623, Dormer to Curzon, 8.11.20.
135 Ibid.
136 MinFo, *Memoria 1920,* Doc. 11, pp. 10–11.
137 The government had brought a similar suit at the Federal Court of
    Cassation against VOC on 24 April 1921, using similar arguments as
    with CDC (cf. MinFo, *Memoria 1921,* Vol. 1, Doc. 12, pp. 12–13).
138 MinFo, *Memoria 1922,* Vol. 1, Docs. 4–9, pp. 8–18.
139 DS 831.6363/Lago Pet. Corp./1, Fred H. Kay (compiler), 'History of
    Petroleum Concessions owned by Lago Petroleum Corporation in
    Venezuela', April 1928.
140 AHMSGPRCP Sept. 1923, José A. Domínguez to E. Urdaneta Maya,
    29.9.23.
141 ADCOTHMEM, 'Traspasos, 1927', Vol. 7.
142 AGTC Enero–Mayo 1921, Gómez (Juan C.) to Gómez, 5.3.21.
143 AHMSGPRCPU, Márquez Bustillos to Urdaneta Maya, 30.12.20.
144 AGTC Abril–Junio 1920, Gómez (José V.) to Torres, 28.7.20.
145 Ibid.
146 AGTC Ago–Dic. 1920, Gómez (José V.) to Torres, 1.12.20.
147 AGTC Enero–Mayo 1921, Gómez (José V.) to Torres, 20.3.21.
148 MinFo, *Memoria 1924,* A.A. Sobalbarro, 'Lago Petroleum Corp. Informe',
    and DS 831.6363/Lago Pet. Corp./1, Kay, 'History of Petroleum'.
149 AGTC Enero–Mayo 1921, Carlos Delfino to Torres, 26.4.21; and
    ADCOTHMEM, 'Historial', Files 13,005–14; 13,040–9; 13,052–4;
    13,063–70 and 13,077–82.
150 AGTC Enero–Mayo 1921, Gómez (José V.) to Torres, 9.5.21.
151 AGTC Enero–Mayo 1921, Gómez (José V.) to Torres, 25.5.21.
152 Raúl Capriles, for example, was assistant to Dr Ezequiel Vivas

(Secretary-General to Gómez) in Maracay and remained so until the mid-1920s.
153 AGMSGPRCS Marzo–Abril 1921, Abraham Capriles to Raúl Capriles, 8.4.21.
154 AHMSGPRCS Julio–Ago 1921, M.J. Capriles to Raúl Capriles, 19.8.21; and Ibid., Abraham Capriles to Raúl Capriles, 28.7.21.
155 AHMSGPRCS Sept.–Oct 1921, A.S. Capriles to Raúl Capriles, 7.10.21.
156 AHMSGPRCS Julio–Ago 1921, M.V. González Rincones to Urdaneta Maya, 17.8.21.
157 Ibid.
158 MinFo, *Memoria 1920,* Exposición, p. xvi.
159 DS 831.6363/61, McGoodwin to Secretary of State, 9.5.21.
160 Ibid.
161 DS 831.6363/61 McGoodwin to Secretary of State, 27.5.21. Enclosure 'Memorándum', undated.
162 Ibid.
163 AGTC Junio–Dic. 1921, and AHMSGPRCS 1–30 Abril 1919 *(sic),* 'Memorándum', undated.
164 MinFo, *Memoria 1920,* Exposición, pp. ix–xii.
165 Ibid., pp. xii–xiii.
166 AGTCOP 7, Torres to Gómez, 27.4.21.
167 AHMSGPRCP 1–30 Abril 1921, Márquez Bustillos to Urdaneta Maya, 27.4.21.
168 AGTC Junio–Dic. 1921, and AHMSGPRCS 1–30 Abril 1919 *(sic),* 'Memorándum', unsigned and undated.
169 AHMSGPRCP 1–30 Abril 1921, Márquez Bustillos to Urdaneta Maya, 27.4.21; and DS 831.6363/61, McGoodwin to Secretary of State, 27.5.21.
170 DS 831.6363/61, McGoodwin to Secretary of State, 27.5.21.
171 Ibid.
172 Ibid.
173 Ibid.
174 Ibid.
175 DS 831.6363/79, J.C. White to Secretary of State, 17.10.21. Enclosure, Translation, Fred H. Kay, 'Solicitud', undated.
176 Ibid.
177 AGTCOP 9, Torres to Gómez, 2.4.22; and 'Los Memorales de Gumersindo Torres', *BAHM,* 2:9(Nov.–Dic. 1960), 157–65.
178 MinFo, *Memoria 1921,* Exposición, p.ix.
179 DS 831.6363/79, White to Secretary of State, 17.10.21.
180 AGTCOP 9, Torres to Gómez, 30.3.22.
181 AGTCOP 9, Torres to Gómez, 2.4.22.
182 'Proyecto de Ley de Impuestos al Petróleo', *Boletín del Ministerio de Fomento,* 2:16(Enero 1922), 721–8.
183 AGTCOP 9, Torres to Gómez, 2.4.22.
184 Ibid.
185 Ibid.
186 Ibid.
187 The taxes which the companies considered confiscatory increased from

Mexican $0.10 to 0.21 in 1921 (cf. Lorenzo Meyer, *México y Estados Unidos en el conflicto petrolero, 1917–1942* (México, El Colegio de México, 1968); and U.S. Senate, *American petroleum interests in foreign countries*, Hearings before a Special Committee Investigating Petroleum Resources, 79 Cong. 1 Sess., 1946).

188 Cf. U.S. Senate, 'Diplomatic correspondence with Colombia in connection with the Treaty of 1914 and certain oil concessions', *Senate Document No.64*, 68 Cong. 1 Sess., 1924; and A.H. Redfield, 'Our petroleum diplomacy in Latin America' (Ph.D. Diss.. The American University, 1942).

189 Cf. U.S. Senate, 'Diplomatic correspondence with Colombia'.

190 George S. Gibb and E.H. Knowlton, *The resurgent years, 1911–1927* (New York, Harper and Bros., 1956), p. 306.

191 AMF, 'Estudio comparativo sobre los impuestos proyectados sobre el petróleo', undated.

192 AGTCOP 9, Torres to Gómez, 30.3.22.

193 'Proyecto de Ley – Ministerio de Fomento', *Boletín del Ministerio de Fomento*, 2:19 (Abril 1922), 862–75.

194 Ibid., p. 863.

195 *DDCS*, Mes 1 (1), 23.4.22, 'Comisión Permanente de la Cámara sobre Fomento', p. 8.

196 ADCOTHMEM, 'Historial', Files 9,760, 14,701 and 14,271.

197 AGTC Enero–Mayo 1922, Torres to Gómez, 29.4.22.

198 AGTCOP 9, Torres, 'Al margen de las observaciones sobre un proyecto de Ley de Hidrocarburos', 10.5.22.

199 Ibid.

200 Ibid.

201 AGTCOP 9, Torres to Gómez, 10.5.22.

202 Lecuna, *Historiador*, Lecuna and H. Pérez Dupuy to Gómez, 18.5.22, pp. 12–13 (p. 12).

203 Ibid., p. 12.

204 Ibid., p. 12.

205 AGTCOP 9, Torres to Gómez, 29.5.22.

206 Ibid.

207 Arcaya, *Memorias*, pp. 168–9.

208 Article 43 of the law and article 22 of Torres' proposal gave the government the right to demand 'special advantages for the Nation in tax matters'.

209 On 12 December 1922 the government agreed to confirm CPC's title for the sum of Bs.10 million. In July 1925 VOC paid Bs.10 million to the government to settle a dispute over its oil concessions (cf. MinFo, *Memoria 1922*, pp. 8–18; and MinFo, *Memoria 1925*, pp. 40–1).

210 Rufino González Miranda, *Estudios acerca del régimen legal del petróleo en Venezuela* (Caracas, UCV, 1958), pp. 490–1.

211 Charles Issawi and Mohammed Yeganeh, *The economics of Middle Eastern oil* (London; Faber and Faber, 1963), Table 34, p.114. The period in question incorporates the added taxation levied by the 1943 Oil Law. According to Parra, between 1917 and 1942, the government received 23 per cent of the companies' gross receipts in taxes. See Alirio Parra, *La*

industria petrolera y sus obligaciones fiscales en Venezuela (Caracas, Primer Congreso Venezolano del Petróleo, 1962).
212 Issawi and Yeganeh, *The economics of Middle Eastern oil*, Table 32, p. 109. Total receipts for Venezuela were $946 million compared to $510 million for the Middle East.
213 Ibid., Table 32, p. 109.
214 Ibid., Table 34, p. 114.
215 Ibid., Table 34, p. 114.
216 DS 831.6363/106, Willis C. Cook to Secretary of State, 22.6.22.
217 Ibid.
218 AHMSGPRCP 1–15 Junio 1922, McKay to Gómez, 15.6.22.
219 Henry C. Morris, 'Fomento de la industria petrolera en las Américas', *Boleín del Ministerio de Fomento* (Segunda Epoca), 2:21 (Junio 1922), 943–7.
220 AGTCOP 9, Torres to Gómez, 12.6.22.
221 AHMSGPRCP 16–30 Abril 1927, V. Pérez Soto to Gómez, 25.5.27.
222 MinFo, *Memoria 1922*, pp. 81–7.
223 Ibid., Exposición, p. vi.

3. Oil Companies and Finance

1 Ministerio de Minas e Hidrocarburos, *Petróleo y otros datos estadísticos* (Caracas 1964), pp. 138–9.
2 Cf. FO 371/11201, Seeds to A. Chamberlain, 12.1.26; FO 371/13558, Sir E. Howard to Chamberlain, 1.10.29; and U.S. Senate, *American petroleum interests in foreign countries*, Hearings before a Special Committee Investigating Petroleum Resources, 79 Cong. 1 Sess., 1946, p. 337.
3 ADCOTHMEM, 'Traspasos, 1926', Vol.3.
4 ADCOTHMEM, 'Traspasos, 1929', Vol.4.
5 ADCOTHMEM, 'Traspasos, 1934'. Brandt retained his 2.5 percent royalty.
6 Walter R. Skinner, *The Oil and Petroleum Manual* (London, 1928).
7 AHMSGPRCS 1–15 Enero 1932, O. Kerdel to Rafael Requena, 11.1.32.
8 Cf. Skinner *Oil Manual* (1935).
9 AHMSGPRCP 1–10 Julio 1930JVG, North Venezuelan Petroleum Co. Ltd to J.B. Pérez, 4.7.30.
10 *The South American Journal*, CXII:1 (2 July 1932).
11 ADCOTHMEM, 'Traspasos, 1927', Vol.5.
12 ADCOTHMEM, 'Traspasos, 1930', Vol.3.
13 ADCOTHMEM, 'Traspasos, 1926', Vol.5.
14 ADCOTHMEM, 'Traspasos, 1926', Vol.5.
15 ADCOTHMEM, 'Traspasos, 1927', Vol.7.
16 ADCOTHMEM, 'Traspasos, 1933–32' *(sic)*.
17 ADCOTHMEM, 'Traspasos, 1934'.
18 ADCOTHMEM, 'Traspasos, 1934'.
19 AHMSGPRCP 16–30 Abril 1923, David Rodríguez to Gómez, 30.3.23.
20 AHMSGPRCP 1–10 Ago 1924, F. Conde García to Gómez, 7.8.24.
21 AHMSGPRCP 1–14 Mayo 1926, M.S.Briceño to Gómez, 5.5.26.
22 AHMSGPRCP 16–30 Junio 1932, José León M. to Gómez, 18.6.32.

23 AHMSGPRCS Mayo 1926, Antonio Alamo to F. Baptista Galindo, 5.5.26.
24 AHMSGPRCP 22–31 Oct. 1924, Rafael A. Hermoso to Gómez, 31.10.24. There is no evidence that this deal went through.
25 AHMSGPRCP 21–31 Junio 1930JVG, Guillermo Paúl to Sixto Tovar, 13.6.30.
26 AHMSGPRCP 15–31 Enero 1930, Guillermo Paúl to Sixto Tovar, 13.6.30.
27 AHMSGPRCP 20–31 Enero 1934, Carmen Lecuna de Blanco to Gómez, 27.1.34.
28 AHMSGPRCS 16–31 Mayo 1934, L. Vallenilla Lanz to E. Urdaneta Carrillo, 20.5.34.
29 AHMSGPRCP 16–31 Mayo 1935, Ubaldo Chiara to Gómez, 30.5.35.
30 Ibid.
31 AHMSGPRCP 1–10 Enero 1932, J.V. López to Gómez, 7.1.32.
32 Ibid.
33 AHMSGPRCS 1–14 Ago 1932, López to Requena, 1.8.32.
34 AHMSGPRCP 11–21 Junio 1935, López to Urdaneta Carrillo, 16.7.35.
35 AHMSGPRCP 1–15 Mayo 1935, López to Gómez, 10.5.35.
36 AHMSGPRCP 1–15 Sept. 1934, Lucio Baldó to Gómez, 15.9.34.
37 AHMSGPRCP 1–9 Junio 1924, P.R. Rincones jun. to Louis B. Wekle, 5.6.24.
38 AHMSGPRCP 1–9 Junio 1924, Fontiveros to Rincones jun., 12.6.24.
39 DS 831.6363/303, Cook to Secretary of State, 26.12.25.
40 AHMSGPRCP 1–20 Abril 1935, G.T. Villegas Pulido to Gómez, 12.4.35.
41 AHMSGPRCP 1–14 Ago 1933, Eduardo E. Santos to Gómez, 9.8.33.
42 AHMSGPRCP 16–30 Abril 1927, Pérez Soto to Gómez, 25.5.27.
43 DS 831.6363/299, Cook to Secretary of State, 27.11.25.
44 DS 831.6363/296, Cook, 'General conditions prevailing in Venezuela', 14.11.25.
45 MinFo, *Memoria 1924*, pp. 16–17.
46 Venezuela, Jurado de Responsabilidad Civil y Administrativa (VJRCA), *Sentencias* (Caracas, Imprenta Nacional, 1946), Vol.4, 'Antonio Alamo', p. 156.
47 Ibid.
48 AHMSGPRCS 15–31 Marzo 1935, Alamo, 'Memorándum para el señor General Juan Vicente Gómez, Presidente de los EE. UU. de Venezuela', 23.2.35.
49 AHMSGPRCS 15–31 Marzo 1935, Alamo to Urdaneta Maya, 15.3.35.
50 Ibid.
51 DS 831.6363/316, Wainwright Abbott to Secretary of State, 13.3.26. Enclosure No.2, 'Agreement', 4.2.26.
52 DS 831.6363/376, C. van H. Engert to Secretary of State, 8.3.28.
53 DS 831.6363/376, Engert to Secretary of State, 28.3.28.
54 VJRCA, *Sentencias*, Vol.1 'Adolfo Bueno', p.105; and Rómulo Betancourt, *Venezuela. Política y petróleo*, 2nd edn (Caracas, Editorial Senderos, 1967), p. 67.
55 DS 831.6363/Falcón Pet. Corp./2, Engert to Secretary of State, 2.10.29.

56 VJRCA, *Sentencias*, Vol. 1, 'Adolfo Bueno', p. 105.
57 PA, W.G. Beavan, 'Mission to Venezuela. Report No.5', 4–7.8.23.
58 AHMSGPRCP 1–14 Oct. 1923, F.A. Colmenares Pacheco to Gómez, 31.10.23.
59 AHMSGPRCP 1–10 Ago 1924, Juan E. Paris to Gómez, 10.8.24.
60 AHMSGPRCP 12–31 Marzo 1931, 'Extracto del contrato celebrado entre Juan E. Paris hijo, como apoderado de R. Isava Núñez, y la Orinoco Oil Company', 22.10.22.
61 Ibid.
62 AHMSGPRCP 1–6 Abril 1925, Gómez (Santos M.) to Gómez, 4.4.25.
63 AHMSGPRCP 10–14 Junio 1928, H.L. Leyva to Tovar, 1.6.28.
64 Ibid.
65 AGTC Enero–Julio 1930, Octavio Calcaño Vethancourt to Torres, 26.4.30.
66 ADCOTHMEM, 'Historial', Files 8,680–8; 9,308–15; 12,546–54; 12,556–76; 12,620–35; 12,832–5; 14,735; 15,990; 16,004; 16,064; 16,076; 16,230; and 16,324.
67 MinFo, *Memoria 1924*, Doc.29, pp. 70–1.
68 'Venezuelan Petroleum Developments', *Petroleum Times* (London), 14:340 (11 July 1925), 51–2, p. 52.
69 *Boletín del Petróleo* (Caracas), 1:1 (1 Marzo 1925), p. 25.
70 DS 831.6363/249, J. Webb Benton to Secretary of State, 30.10.24.
71 DS 831.6363/182, F.C. Chabot to Secretary of State, 21.5.24.
72 AHMSGPRCP 16–30 Nov. 1925, Torres to Gómez, 20.11.25. When the concessions were sold Torres received Bs.39,520.
73 AHMSGPRCP 1–6 Abril 1925, A.J. Ramírez Román to Gómez, 16.4.25. The concessions were later sold to the Venezuelan Pantepec Oil Co. for Bs.25,825. Ramírez received Bs.5,837.3 and Juancho Gómez Bs.989.55.
74 DS 831.6363/AT6, Abbott to Secretary of State, 24.11.24.
75 DS 831.6363/242, Chabot to Secretary of State, 23.8.24.
76 Cf. B.S. McBeth, *Royal Dutch-Shell vs. Venezuela* (Oxford Microfilm Publications, 1982), pp. 182–215.
77 *Gaceta Oficial*, No. 16,307, 12.9.27.
78 ADCOTHMEM, 'Traspasos, 1927', Vol.5.
79 AHMSGPRCS Oct. 1922, P.R. Rincones to Urdaneta Maya, 7.9.22.
80 PA, Fields to Ryder, 17.9.23.
81 PA, 'Plan advanced by the undersigned for the organization of the American Venezuelan Development Corporation', undated.
82 PA, Fields to Ryder, 17.9.23.
83 Ibid.
84 AHMSGPRCP 1–15 Abril 1923, J.A. Coronil to Gómez, 7.4.23.
85 PA, Fields to Ryder, 17.9.23.
86 AHMSGPRCS Mayo 1923, Arcaya to Urdaneta Maya, 2.5.23.
87 DS 831.6363/134, Cook to Secretary of State, 14.5.23.
88 PA, Ryder to Lord Cowdray, 17.9.23.
89 Skinner, *Oil Manual*, (1925), p. 111.
90 AHMSGPRCP 15–31 Sept. 1924, Delfino to Gómez, 23.9.24.
91 VJRCA, *Sentencias*, Vol. 2, 'Josefina Revenga de Tinoco', p. 183.
92 Ibid. The shares and debentures were distributed as follows: Josefina

Gómez Revenga de Olavarría: 533 shares and 30 debentures; José
Vicente Gómez Revenga: 533 shares and 30 debentures; and Alí Gómez
Revenga: 534 shares and 29 debentures.
93 Ibid., and AHM, Gómez (José V.) to Gómez, 24.7.29.
94 AGTC Julio–Dic. 1931, Torres to R. Cayama Martínez, 31.7.31.
95 AHMSGPRCP 11–20 Marzo 1928, Pedro F. Lehman to Gómez, 16.3.28.
96 AHMSGPRCP 1–11 Julio 1926, Guillermo Elizondo to Gómez, 2.7.26.
97 AHM, Gómez to The Venezuelan Gulf Oil Co., 18.3.30.
98 AHMSGPRCP 1–15 Julio 1925, W. Rogers to Gómez, 4.7.24.
99 AHMSGPRCP 1–12 Feb. 1927, Alamo to Gómez, 7.2.27.
100 AHMSGPRCP 15–21 Sept. 1927, Lorenzo Mercado to Gómez, 13.12.27.
101 AHMSGPRCP 11–19 Dic. 1927, Mercado to Gómez, 13.12.27.
102 AHMSGPRCP 1–9 Junio 1928, José A. Tagliaferro to Gómez, 6.6.28.
103 AGTCOP 9, Torres to Gómez, 10.5.22.
104 AHM, Compañía Venezolana de Petróleo, *Estatutos de la Compañía
Anónima 'Compañía Venezolana de Petróleo'* (Caracas, NP, 1923).
105 DS 831.6363/138, Cook to Secretary of State, 3.7.23.
106 FO 371/9638, A.P. Bennett to Foreign Office, 28.4.24.
107 VJRCA, *Sentencias*, Vol.3, 'Cia. Venezolana de Petróleo', p. 272.
108 They nevertheless shared in the profits of the company. González Rincones
and Baldó declared later that their share varied between Bs.450,000 and
500,000. Ramírez's share was larger. (Ibid., 'Declaraciones de Roberto
Ramírez', pp. 312–6; Ibid., 'Declaraciones del Dr R. González Rincones',
p.317.)
109 Ibid., 'Declaraciones de Roberto Ramírez', p. 315.
110 Ibid., p. 312–3.
111 AHMSGPRCP 17–31 Dic. 1923, Rafael Falcón to Gómez, 19.12.23.
112 AHMSGPRCP 1–9 Junio 1924, G. Willet to Gómez, 4.6.24.
113 VJRCA, *Sentencias*, Vol. 3, 'Cia. Venezolana de Petróleo', p.272.
114 Edwin Lieuwen, *Petroleum in Venezuela* (Berkeley, University of California
Press, 1954), pp. 73–4.
115 Roberto Ramírez and Wriglah (*sic*) Waltking, 'Convenio celebrado entre
la Cia. Venezolana de Petróleo y el grupo de industriales por el sr. W.
Waltking', 21.2.24 in DS 831.6363/195, Chabot to Secretary of State,
6.5.24, Enclosure.
116 The purchase price was $3 million (Bs. 15.6 million); payment for option
was $50,000 (Bs.260,000), with $700,000 (Bs.3,640,000) being paid when
option was exercised, and the balance to be held in CVP's stock (up to a
maximum of a quarter of the company's equity). (DS 831.6363/160, Cook
to Secretary of State, 7.3.24.)
117 DS 831.6363/160, Cook to Secretary of State, 7.3.24.
118 Ibid.
119 DS 831.6363/169, Chabot to Secretary of State, 5.4.24.
120 DS 831.6363/172, Chabot to Secretary of State, 13.4.24.
121 DS 831.6363/178, Chabot to Secretary of State, 14.4.24.
122 DS 831.6363/171, Chabot to Secretary of State, 5.4.24.
123 DS 831.6363/180, Chabot to Secretary of State, 26.4.24.
124 'Alemanes é Ingleses y el Petróleo Venezolano', *BAHM*, 12:68 (Enero–

Junio 1971), 131–75, Wilhelm Waltking to D. Lossada Díaz, 12.5.24.
125 DS 831.6363/196, Chabot to Secretary of State, 9.5.24.
126 DS 831.6363/206, Chabot to Secretary of State, 7.6.24.
127 DS 831.6363/211, Chabot to Secretary of State, 14.6.24, Enclosure, R.E. Harwicke to H.J. Stone, 7.6.24; DS 831.6363/224, Chabot to Secretary of State, 24.7.24; and AHMSGPRCP 1–9 Junio 1924, Alamo to Gómez, 6.6.24.
128 DS 831.6363/253, Cook to Secretary of State, 9.1.25; and DS 831.6363/264, Cook to Secretary of State, 3.3.25.
129 DS 831.6363/324, 'Activities of American Commercial Interests', 1.6.26.
130 Ibid.
131 Ibid.
132 'Diógenes Escalante propone la participación venezolana en la explotación petrolera, 1926', *BAHM*, 13:70, 347–52, 'Memorándum de Diógenes Escalante (Para el Doctor Gil Fortoul)', 1.5.26, pp. 348–51 (p. 348).
133 Ibid., p. 349.
134 Ibid., p. 350.
135 Ibid., Escalante to Gómez, 23.7.26, pp. 351–2. Gil Fortoul managed to transfer his gold and diamond mining titles as well as water and railroad interests in Bolívar State to a British mining syndicate, headed by George F. Naphen, and closely associated with the Consolidated Goldfields of South Africa.
136 The company was managed by George F. Naphen, who also controlled the Creole Syndicate. Both companies in turn were owned by the New York bankers Blair and Co. (DS 831.6363/South American Oil and Development Corp./1, Engert to Secretary of State, 15.2.28; DS 831.6363/South American Oil and Development Corp./4, Engert to Secretary of State, 1.5.28; and Interview: Antonio Aranguren Fonseca, 14.9.77).
137 DS 831.6363/South American Oil and Development Corp./2, Engert to Secretary of State, 10.3.28, Enclosure, Clinton D. Winant and J. Gil Fortoul, 'Acuerdo', 15.10.27. A supplementary agreement was entered into on 8 November (Ibid. and AHMSGPRCP 16–31 Oct. 1927, José Gil Fortoul to Gómez, 16.10.27). Both Aranguren and Heyden received a royalty payment for their efforts. (Interview: Antonio Aranguren Fonseca, 14.9.77.)
138 AGTAPM 1930, Roberto Ramírez, 'Memorándum para el sr. Gumersindo Torres, sobre la negociación con la South American Oil and Development Corporation', undated.
139 AHMSGPRCP 21–31 Marzo 1928, Gil Fortoul to Gómez, 27.3.28.
140 AGTAPM 1930, Ramírez, 'Memorándum'.
141 AGTAPM 1930, Ramírez, R. González Rincones and Baldó to Gómez, 31.10.27.
142 Ibid.
143 Made up of 25 per cent to CVP and 15 per cent to Gil Fortoul.
144 AGTAPM 1930, Ramírez *et al.* to Gómez, 31.10.27.
145 Ibid.
146 AGTAPM 1930, Ramírez *et al.* to Gómez, 13.12.27.

147 DS 831.6363/South American Oil and Development Corp./4, Engert to Secretary of State, 1.5.28. Although the Creole Syndicate held control of the company, in a secret agreement CVP retained 62.5 percent of the equity of the South American Oil and Development Corp. (AGTC Enero–Junio 1931, Ramírez to Luis A. Perez, 17.4.31; and AHMSGPRCS 1–28 Feb. 1934, González Rincones to Urdaneta Carrillo, 28.2.34).
148 Skinner *Oil Manual* (1932).
149 AAAC 1, Nestor (Luis Pérez) to Antonio (Aranguren), 22.6.28.
150 AHMSGPRCP 11–20 Nov. 1928, 'Memorándum del Doctor Gil Fortoul', 20.11.28.
151 Ibid.
152 AHMSGPRCP 1–9 Ago 1928, Gil Fortoul to Gómez, 4.8.28.
153 AHMSGPRCP 1–10 Dic. 1928, L.F. Calvani to Gómez, 7.12.28; and AGTAPM 1930, 'Contrato, José Gil Fortoul and Edward Herbert Keeling', undated.
154 AGTAPM 1930, Ramírez, 'Memorándum'.
155 Ibid.
156 AGTC Sept–Dic. 1929, José Santiago Rodríguez, 'Memorándum del Dr J.S. Rodríguez para el Dr Gil Fortoul', 26.2.29.
157 Ibid.
158 AGTAPM 1930, Ramírez, 'Memorándum'.
159 AHMSGPRCP 1–11 Marzo 1930, Gil Fortoul to Gómez, 5.3.30.
160 AGTAPM 1930, Ramírez, 'Memorándum'.
161 AGTAPM 1930, Letter, unsigned, to Gil Fortoul and Ramírez, 31.1.30.
162 AGTC Enero–Junio 1931, Ramírez to Pérez (L.A.), 7.4.31.
163 Ibid.
164 AGTC Enero–Junio 1931, Tomás Liscano to Torres, 24.6.31.
165 AHMSGPRCP Junio 10–14 1928, José Ignacio Cárdenas to Gómez, 13.6.28.
166 AGTC Sept–Dic. 1929, Pérez (L.A.), 'Memorándum para el Dr G. Torres', 17.10.29.
167 VJRCA, *Sentencias*, Vol.3, 'Cia. Venezolana de Petróleo' p. 272. In addition CVP paid for the machinery in Gómez's cosmetic factory of El Prado in Maracay, as well as for the fittings and construction material for his Hotel Miramar, also in Maracay; it also acquired for Gómez the Bramón, Guayabitas and La Horqueta estates.

4 National and Local Effects of the Oil Industry

1 Domingo Alberto Rangel, *Capital y Desarrollo. El Rey Petróleo* (2 vols., Caracas, UCV, 1970), Vol. 2, p. 133.
2 Domingo Alberto Rangel, *Los Andinos en el poder* (Caracas, NP, 1964), pp. 185–8.
3 Esteban Roldán Oliarte, *El General Juan Vicente Gómez. Venezuela de cerca* (México, Imprenta Mundial, 1933), p. 266.
4 H. Brett, 'Venezuela', *Supplement to Commerce Reports*, No.51a (1920), p. 8.
5 S.J. Fletcher, 'Venezuela', *Supplement to Commerce Reports*, No.13 (1922), p. 20.

242   *Notes to pp. 110–17*

6 H. Bancroft-Livingston, *Report on the economic and financial conditions in Venezuela* (London, Dept of Overseas Trade Series, HMSO, 1927).
7 E.D. Crab, 'La Oficina Comercial de los Estados Unidos en Caracas', *BCCC*, 17:170(1.1.28), 3951.
8 U.S. Tariff Commission, *The foreign trade of Latin America. A report on the trade of Latin America with special reference to trade with the United States under the provision of title III, part II, section 332 of the Tariff Act of 1930* (Washington; USGPO, 1942), Appendix B, Tables III & IV, pp. 69–70.
9 Trade would have been greater had her high tariff barrier been reduced (AHMSGPRCP 1–28 Feb. 1920, N. Veloz Goiticoa, 'Reseña sinóptica sobre la segunda conferencia financiera Panamericana', undated.)
10 Cf. Manuel R. Egaña, *Tres décadas de producción petrolera* (Caracas, Tip. Americana, 1947). In Mexico oil revenues only accounted for a third of total revenues in 1922; for the period 1917–35 they averaged 14.5 percent of total revenues. (Cf. Lorenzo Meyer, *México y Estados Unidos en el conflicto petrolero, 1917–1942* (México, El Colegio de México, 1968), Table 4, p. 31.)
11 It should be noted that part of these revenues were directly attributable to the increase in imported goods as a result of the oil boom during the 1920s.
12 Roldán Oliarte, *El General Gómez*, p.267; and, Egaña, *Tres décadas*, p. 12.
13 'Cuadros que manifiestan la disminución de las deudas de Venezuela en los diez años comprendidos de 1 de enero de 1909 a 31 de diciembre de 1918', *BCCC*, 8:69(1.8.19), 596–600.
14 AHMSGPRCP 1–14 Marzo 1931JVG, A.Iselin and Co., 'A review of the financial and economic situation of Venezuela', New York, 26.12.30.
15 Cf. Armando Córdoba, 'La estructura económica tradicional y el impacto petrolero en Venezuela', *Economía y Ciencias Sociales*, 5:1 (Enero-Marzo 1963), 7–28, Tables 1–4 (p. 21).
16 Cf. María de Lourdes Acedo de Sucre and Carmen Margarita Nones Mendoza, *La Generación Venezolana del 1928.* (Caracas, Editorial Ariel, 1967), p. 74; and, MinFo, Dirección General de Estadística, *Números índices de precios al por mayor (1913–1937)* (Caracas, Lit. y Tip. Casa de Especialidades, 1937).
17 Cf. 'La crisis actual', *BCCC*, 9:85 (Dic. 1920), 891; 'Situación Mercantil', *BCCC*, 11:99(1.2.22), 1432–4; and 'Situación Mercantil', *BCCC*, 15:155(1.10.26), 3421–36.
18 'Situación Mercantil', *BCCC*, 16:163 (1.6.27), 3723–30 (p. 3723).
19 FO 371/12063, W. O'Reilly to Chamberlain, 11.7.27.
20 Cf. Alberto Adriani, 'La crisis, los cambios y nosotros', *Cultura Venezolana*, 14:112 (Mayo–Junio 1931), 87–112; and Córdoba, 'La estructura tradicional'.
21 Cf. E. Jeffrey Stann, 'Caracas, Venezuela 1891–1936: a study of urban growth' (Ph.D. Diss., Vanderbilt University, 1975), Table II-8, f. 56.
22 Ibid., Tables II-5 and II-7, ff. 43 and 45 respectively.
23 Claudio Urrutía, 'Apuntes sobre la industria petrolera en Venezuela y su influencia en la vida económica del país', *Boletín del Ministerio de Fomento*, Número Extraordinario (28.7.34.), pp. 3039–49.
24 U.S. Senate, *American petroleum interests in foreign countries*, Hearings before a

Special Committee Investigating Petroleum Resources, 79 Cong. 1 Sess., 1946, p. 239.
25 U.S. Federal Trade Commission, *Economic report on the investigation of coffee prices. Summary and conclusions* (Washington; USGPO, 1954), Appendix E, Table E-1, p. 520.
26 Hugo N. Guardia, *Recopilación de Estadística cafetera* (Caracas; Publicaciones del Instituto Nacional del Café, 1943), pp. 5–8.
27 Alberto Adriani, *Labor venezolanista* (Caracas; Tip. La Nación, 1937), p. 82.
28 G. Delgado Palacios, *Contribución al estudio del café en Venezuela* (Caracas; Tip. El Cojo, 1895).
29 Cf. 'Apuntes sobre la situación cafetera venezolana', *Boletín del Banco Central de Venezuela*, 6:24 (Feb. 1947), 8–10.
30 Angel Biagini, 'En torno de la prima del café', *Revista del Instituto Nacional del Café*, 1:1 (Ago 1939), 63–70.
31 Oscar Linares, 'Apreciaciones sobre la producción de café en Venezuela comparada con la de Colombia', *BCCC*, 14:143 (1.10.25), 3016–18.
32 Ibid.
33 'Apuntes sobre la situación cafetera venezolana', *Boletín del Banco Central de Venezuela*, 6:24 (Feb. 1947), 8–10.
34 N. Veloz Goiticoa, *Venezuela* (Caracas, Tip. Central, 1919).
35 Charles J. Dean, 'Commerce and industrial development in Venezuela', *Trade Information Bulletin*, No. 783 (1931).
36 'Situación mercantil', *BCCC*, 16:164 (1.7.27), 3763–9.
37 Hermán Nass, *El crédito agrícola en Venezuela* (Caracas, Cuadernos Verdes No. 14, Editorial Crisol, 1945).
38 Ramón León, 'Consideraciones sobre el café', *Revista del Instituto Nacional del Café*, 1:1 (Ago 1939), 71–6.
39 Jaime Henao Jaramillo, *La caficultura y la economía nacional* (Caracas, Cuadernos Verdes No. 84, Tip. El Compás, 1950).
40 Federico Brito Figueroa, *Venezuela siglo XX* (La Habana, Casa de las Américas, 1967), p. 41.
41 Walter Dupouy 'El petróleo y las tierras agro-pecuarias', *El Farol*, 10:111 (Ago. 1948), 2–9 (p. 7).
42 'Petróleo', *BCCC*, 13:124 (1.3.24), 2331.
43 'La agricultura ante el petróleo', *BCCC*, 16:162 (1.5.27), 3714–5.
44 'Situación mercantil', *BCCC*, 16:167 (1.10.27), 3838–43 (p. 3839).
45 Brito, *Venezuela siglo XX*; Rangel, *Capital y Desarrollo;* and Rodolfo Quintero, *La cultura del petróleo* (Caracas, Colección Esquema Gráfica Universitaria C.A., 1968).
46 MinFo, Dirección General de Estadística, *Estadística del petróleo para los años 1936, 1937, 1938* (Caracas, 1940), p. vii.
47 Ibid.
48 P.L. Bell, 'Venezuela; a commercial and industrial handbook' *Special Agents Series No. 212,* (1922).
49 The indians were kept as quasi-slaves and bought as children for Bs.40 (cf. Brito, *Venezuela siglo XX*, p. 86).
50 Venezuela, *El General Juan Vicente Gómez. Documentos para la historia de su*

Gobierno (compiled by Luis Correa) (Caracas, Lit. del Comercio, 1925), Gómez to State Presidents, 21.9.11, p. 172.

51 T. Ifor Rees, 'Venezuela – Report for the year 1913–14 on the trade of Venezuela and the Consular District of Caracas', *PP*, LXXV (1915), 849–81 (p. 859).

52 Ibid.

53 MinFo, *Memoria 1917.*

54 MinFo, Dirección de Estadística, Estado Zulia, *Censo industrial, comercial y empresas que prestan servicios – 1936* (Caracas, Tip. Garrido, 1939), Tables 7, pp. 89, 187 and 263 respectively.

55 Rangel, *Capital y Desarrollo*, p. 231.

56 F. Mieres, 'Los efectos de la explotación petrolera sobre la agricultura de Venezuela' in Héctor Malavé Mata, *Petróleo y Desarrollo económico de Venezuela* (La Habana, Publicaciones Económicas, 1964), 343–71 (p. 352).

57 Brito, *Venezuela siglo XX*, p. 101.

58 Cf. Rodolfo Luzardo, *Venezuela: Business and finances* (Englewood Cliffs, New Jersey, Prentice-Hall, 1957), Table 12, p. 69.

59 U.S. Dept. of Commerce, *Investment in Venezuela* (Washington, USGPO, 1953).

60 Rodolfo Quintero, *Antropología de las ciudades latinoamericanas* (Caracas, UCV, 1965), p. 91.

61 AHMSGPRCP 11–20 Enero 1908, José Ignacio Lares to C. Castro, 16.1.08.

62 AHMSGPRCP 1–15 Abril 1918, Gómez (Santos M.) to Gómez, 2.4.18.

63 A.J. Briceño Parilli, *Las migraciones internas y los municipios petroleros* (Caracas, Tip. ABC, 1947).

64 FO 371/12063, O'Reilly to Chamberlain, 2.3.27.

65 *Oil News*, 26:887 (6.12.29), 573.

66 John Robert Moore, 'The impact of foreign direct investment on an underdeveloped economy. The Venezuelan case' (Ph.D Diss., Cornell University, 1956).

67 Venezuela, Banco Central de Venezuela, *Ingreso Nacional de Venezuela en 1936* (Cracas, Monografías del Banco Central de Venezuela No. 1, Editorial Relámpago, 1949), Table 1, p. 93.

68 Rangel, *Capital y Desarrollo*, Table 21, p. 149.

69 Ibid., p. 247.

70 AHMSGPRCP 20–30 Nov. 1930JVG, Cecilio L. Duchanf, 'Memorándum sobre la situación presente de los hacendados de café', Nov. 1930.

71 AHMSGPRCP 1–10 Enero 1932 *(sic)*, Manager, Banco Agrícola y Pecuario, to Ministerio de Salubridad y Agricultura y Cría, 27.1.31.

72 FO 199/275, J. Keeling to Foreign Office, 13.4.33.

73 AHMSGPRCP 11–19 Junio 1933, S. Itriago Chacín to Gómez, 19.6.33.

74 AHMSGPRCP 16–30 Abril 1934, J.A. González to Gómez, 24.4.34.

75 AHMSGPRCP 16–30 Abril 1934, Ramón Penzini and Carlos M. Iragan to Gómez, 30.4.34.

76 MinRelInt, *Memoria 1934*, Doc. 4, 24.7.34. pp. 6–7.

77 J.V. Gómez, *Mensaje que el ciudadano General J.V. Gómez, Presidente de los EE.UU. de Venezuela, presenta al Congreso Nacional en sus sesiones ordinarias de*

1935 (Caracas, Lit. del Comercio, 1935).
78 AHMSGPRCP 20–31 Oct. 1934 *(sic)*, Manuel Toledo Trujillo *et al.* to Gómez, 26.7.34.
79 AHMSGPRCP 1–14 Nov. 1935, Ramón J. López Rodríguez to Gómez, 4.11.35.
80 Jaime Henao Jaramillo, 'La industria cafetera de Venezuela', *Revista del Instituto Nacional del Café*, 3:10 (Dic. 1941), 35–9 (p. 37).
81 David Lawrence Taylor Knudson, 'Petroleum, Venezuela, and the United States, 1920–1941' (Ph.D. Diss., Michigan State University, 1975), f. 21.
82 Ibid.
83 'Situación mercantil', *BCCC*, 19:203 (Oct. 1930), 4905–8 (p. 4905).
84 AHMSGPRCS 15–30 Sept. 1931, Abel Santos, 'Notas', 25.9.31.
85 AHMSGPRCP 19–30 Sept. 1931, Bernardo Jurado Blanco to Gómez, 26.9.31.
86 AHMSGPRCP 1–15 Nov. 1931, Jurado Blanco 'Memorándum Causa de la baja del bolívar – Necesidad de la nacionalización de los bancos y casas bancarias extranjeras, tal como lo han hecho otros países', 1.11.31.
87 'Situación mercantil', *BCCC*, 20:215 (Oct. 1931), 5263–6 (p. 5263).
88 AHMSGPRCP 11–20 Abril 1932, G.J. Sanabría, 'Memorándum', 13.4.32.
89 AHMSGPRCP 20–31 Ago 1934, Requena to Gómez, 28.8.34.
90 AHMSGPRCP 11–20 Sept. 1932, Arcaya to Gómez, 20.9.32.
91 AHMSGPRCP 21–31 Oct. 1933, Arcaya to Gómez, 24.10.33.
92 Ibid.
93 Ibid.
94 AHMSGPRCP 16–30 Junio 1932, Gustavo Brandt to Gómez, 22.6.32.
95 'La agricultura y la moneda', *BCCC*, 23:248 (July 1934), 6193–5.
96 FO 371/17618, British Legation to Dept of Trade, 30.8.34.
97 Ibid.
98 AHMSGPRCP 1–10 Oct. 1934, Arcaya to Gómez, 2.10.34.
99 Mieres, 'Los efectos de la explotación', p. 367.
100 'Situación mercantil', *BCCC*, 14:144 (1.11.25), 3033–8 (p. 3037).
101 'Progresos de la explotación de petróleo en el Zulia', *BCCC*, 14:136 (1.3.25), 2764.
102 'Situación mercantil', *BCCC*, 15:148 (4.3.26), 3163–7; and *Latin American World*, 7:10 (June 1926).
103 'Situación mercantil', *BCCC*, 15:148 (4.3.26), 3163–7.
104 'Situación mercantil', *BCCC*, 15:146 (1.1.26), 3101–6 (p. 3105).
105 AHMSGPRCP 11–20 Abril 1926, P. León to Gómez, 16.4.26.
106 AHMSGPRCP 1–15 Abril 1916, Alberto Aranguren to Gómez, 7.4.16.
107 AHMSGPRCP 1–31 Mayo 1919, José J. Gabaldón to Gómez, 18.7.19.
108 AHMSGPRCP 1–28 Feb. 1920, Unsigned to Gómez, 2.2.20.
109 AHMSGPRCP 16–31 Marzo 1922, Palmarito Sugar Co., 'Memorándum presentado al señor General Juan Vicente Gómez por la Palmarito Sugar Company', 22.3.22.
110 AHMSGPRCP 1–10 Ago 1924, Albino de J. Medina to Gómez, 1.8.24.
111 AHMCOP 233, Gómez to M. Toro Chimies, 11.2.24.
112 Ibid.

113 AHM, L.F. Heghon and C.M. Crebbs to G.C. Chio *et al.*, 1.6.25.
114 AHMSGPRCP 22–31 Julio 1925, César A. León to Gómez, 26.7.25.
115 FO 199/218, H.A. Hobson to Chamberlain, 17.7.25.
116 FO 199/218, G. Witteveen to VOC, 24.7.25.
117 FO 199/218, Hobson to Chamberlain, 30.7.25.
118 FO 199/218, Hobson to Chamberlain, 26.6.25.
119 FO 199/218, Witteveen to VOC, 17.7.25.
120 FO 199/218, Hobson to Chamberlain, 17.7.25.
121 DS 831.6363/319, Abbot to Secretary of State, 20.3.26.
122 Briceño Parilli, *Las migraciones internas.*
123 Jesús Prieto Soto, *Huellas históricas de Cabimas* (México, Editorial cultura, 1959), p. 33.
124 AHMSGPRCP 1–14 Feb. 1926, A.A. Sobalvarro to Arcaya, 11.2.26.
125 Ibid.
126 Ibid.
127 L.F. Calvani, 'Informe del Inspector Técnico de Minas al Ministerio de Fomento', *BCCC,* 14:142 (1.9.25), 2979–80.
128 'El Zulia en 1926 y en 1929', *BAHM,* 17:90 (Marzo–Abril 1976), 53–77, Isilio Febres Cordero to Gómez, 16.2.26, pp. 57–8.
129 AMF, J.M. Braschi to Alamo, 17.2.26; and Ibid., Alamo to Inspector Fiscal de Hidrocarburos, 18.2.26.
130 AMF, Braschi to Alamo, 22.2.26.
131 AHMSGPRCP 11–20 Marzo 1926, Clementina Romero to Gómez, 15.3.26.
132 Ibid.
133 'Pérez Soto y las compañías petroleras, 1926', *BAHM,* 13:70 (Enero–Feb. 1972), 319–46, V. Pérez Soto to Gómez, 'Memorándum No. 18', 7.7.26, pp. 328–30 (p. 329).
134 Ibid., p. 330.
135 AHMSGPRCP 11–20 Abril 1926, León (P.) to Gómez, 20.4.26.
136 'El Zulia en 1926 y en 1929', *BAHM,* Héctor García Chuecos to Gómez, 21.4.26, pp. 60–1.
137 Ibid., Febres Cordero to Gómez, 16.2.26, pp. 56–7.
138 Ibid.. Febres Cordero to Gómez, 15.5.26, pp. 63–4.
139 'Venezuela en la Liga de las Naciones (1925)', *BAHM,* 13:70 (Enero–Feb. 1972), 207–39, José Ignacio Cárdenas to Gómez, 24.3.24, pp. 213–14 (p. 214); and DS 831.6363/318, A.K. Sloan, 'Political rumours in Maracaibo', 18.3.26.
140 AHMSGPRCP Sept. 1926, Pérez Soto to Baptista Galindo, 25.9.26.
141 AHMSGPRCP 16–31 Oct. 1927, Pérez Soto to Gómez, 24.10.27.
142 'Pérez Soto y las compañías petroleras', *BAHM, 'Memorándum No. 17',* Pérez Soto to Gómez, 5.7.26, p. 328.
143 Ibid.
144 AHMSGPRCP Sept. 1926, Pérez Soto to Baptista Galindo, 25.9.26.
145 'Pérez Soto y las compañías petroleras', *BAHM,* 'Memorándum No. 29', Pérez Soto to Gómez, 22.8.26, pp. 341–3 (p. 341).
146 Ibid., p. 342.
147 MinRelInt, *Memoria 1926,* Pérez Soto to Interior Minister, 23.7.26, Doc.

62, pp. 101—3; and Ibid., Arcaya to Pérez Soto, 5.8.26, p. 103.
148 'Pérez Soto y las compañías petroleras', *BAHM*, 'Memorandum No. 26',
    Pérez Soto to Gómez, 16.6.26, p. 338 and p. 345.
149 AHMSGPRCP 11–19 Marzo 1927, Pérez Soto, 'Memorándum No. 43
    para J.V. Gómez', 16.3.27. The following companies were affected: CDC,
    VOC, Venezuelan Sun Co., Maracaibo Oil Co., and the Orinoco Oil Co.
150 AHMSGPRCP 16–31 Enero 1922, Gómez (Santos M.) to Gómez, 26.1.22.
151 AHMSGPRCP 20–31 Marzo 1927, Pérez Soto, 'Memorándum No. 44
    para el Benemérito General Juan Vicente Gómez', 25.3.27.
152 Zulia, *Memoria y Cuenta, 1926*, Pérez Soto to Interior Minister, 22.7.26,
    pp. 150–1.
153 Ibid., p. 151.
154 Ibid., p. 151.
155 MinRelInt, *Memoria 1926*, Arcaya, 'Resolución', 15.7.26, Doc. 143,
    pp. 305.
156 Zulia, *Memoria y Cuenta, 1926*, Pérez Soto to Arcaya, 22.7.26, pp. 150–1
    (p. 151).
157 Ibid., Leonte Olivo to Lago and VOC, 30.4.27, pp. 115–16.
158 AHMSGPRCP 22–30 Junio 1926, R.A. Mora to Gómez, 26.6.26.
159 MinRelInt, *Memoria 1926*, Arcaya to Alamo, 7.8.26, p. 306.
160 Ibid., Arcaya to Pérez Soto, 24.9.26, p. 307.
161 Zulia, *Memoria y Cuenta, 1926*, A. Gavino to Leonte Olivo, 14.8.26, p. 153.
162 Ibid., De Booy to Secretary-General, 4.12.26, p. 163.
163 Zulia, *Memoria y Cuenta, 1928*, Pedro Pinto S. to Secretary-General,
    20.5.28, 92–3 (p. 92).
164 Ibid., Olivo to Venezuelan Gulf Oil Co., 24.5.28, pp. 94–5; and Ibid.,
    Crebbs to Olivo, 12.6.28, pp. 95–6.
165 'Pérez Soto y las compañías petroleras', *BAHM*, 'Memorándum No. 18',
    Pérez Soto to Gómez, 7.7.26, 328–30 (p. 329).
166 Zulia, *Memoria y Cuenta, 1926*, B.Th. W. van Hasselt to Concejo
    Municipal del Distrito Bolívar, 17.7.26, p. 171.
167 This was illegal as under article 4 of the Ley Orgánica del Poder
    Municipal he only had the power to call substitutes. Alternatively a new
    State Constituent Assembly could appoint a new council.
168 Cf. José Rafael Mendoza, *Juicio propuesto por el Concejo Municipal del Distrito
    Bolívar del Estado Zulia contra The Venezuelan Oil Concessions Limited por
    nulidad del contrato de 13 de octubre de 1926 sobre ejidos en Lagunillas, Cabimas, y
    demás Municipios del Distrito Bolívar. (Demanda, contrato, informes, sentencias de
    Primera Instancia)* (Caracas, Impresores Unidos, 1945).
169 Zulia, *Memoria y Cuenta, 1926*, G.T. Villegas Pulido, 'Informe', 3.3.26, pp.
    167–9; and Ibid., Arcaya to Pérez Soto, 7.7.26, pp. 170–1.
170 AHMSGPRCP Ago 1926, Pérez Soto to Arcaya 21.8.26.
171 Ibid.
172 Ibid.
173 Ibid.
174 AHMSGPRCP Ago 1926, Pérez Soto to Arcaya, 21.8.26.
175 Zulia, *Memoria y Cuenta, 1926*, Carlos Dupuy to van Hasselt, 15.7.26,
    pp. 172–4 (p. 173).

248   *Notes to pp. 148–52*

176 AHMSGPRCP 11–20 Feb. 1931JVG *(sic)*, 'Contrato. Concejo Municipal del Distrito Bolívar del Estado Zulia (Presidente Carlos Dupuy Briceño) and Venezuelan Oil Concessions Limited (Rep. Barthold Theodor Whilhelm van Hasselt)', 13.10.26.
177 AHMSGPRCP 16–30 Sept. 1926, Pérez Soto, 'Memorándum No. 33 para el Benemérito General Juan Vicente Gómez', 17.9.26.
178 AHMSGPRCP 20–31 Marzo, 1927, Pérez Soto, 'Memorándum No. 44 para el Benemérito General Juan Vicente Gómez', 25.3.27.
179 Zulia, *Memoria y Cuenta, 1926,* Howland Bancroft to Pérez Soto, 12.8.26, p. 157; and Ibid., Crebbs to Pérez Soto, 14.8.26, p. 158.
180 Ibid., Bancroft to Pérez Soto, 12.8.26, p. 157.
181 AHMSGPRCS Sept. 1923 *(sic)*, Pérez Soto to Baptista Galindo, 25.9.26.
182 Zulia, *Memoria y Cuenta, 1926,* Pérez Soto to Arcaya, 17.8.26, pp. 160–1 (p. 161).
183 Ibid., Pérez Soto to Crebbs, 17.8.26, pp. 158–9.
184 AHM, Pérez Soto to Gómez, 1.11.26.
185 Ibid.
186 Zulia, *Memoria y Cuenta, 1927,* Sobalvarro to Arcaya, 13.11.26, p. 125; and Ibid., R.W. Hardwicke to Arcaya, 15.11.26, pp. 126–7.
187 Ibid., Alamo to Sobalvarro, 7.1.27, p. 123.
188 Ibid., Arcaya to Pérez Soto, 11.6.27, p. 130.
189 Ibid., Olivo, 'Resolución', 15.7.27, pp. 134–6.
190 AHMSGPRCP 1–15 Abril 1927, Pérez Soto, 'Memorándum No. 45 para el Benemérito General Juan Vicente Gómez, 1.4.27.
191 Ibid.
192 Ibid.
193 AHMSGPRCP 11–20 Marzo 1928, Pérez Soto, 'Memorándum No. 70 para el Benemérito General Juan Vicente Gómez', undated.
194 AHMSGPRCP 1–10 Marzo 1928, P.M. Reyes to Gómez, 9.3.28.
195 *El Republicano* (newspaper of the Partido Republicano Venezolano, Panamá), 'La Política petrolera de Estados Unidos', and *La Vanguardia Liberal* (Bucaramanga) in AHMSGPRCP 1–10 Oct. 1928, Pérez Soto to Gómez, 10.10.28. The republic would incorporate Zulia and the Colombian Departments of Magdalena and Santander.
196 AHMSGPRCP 1–9 Ago 1928, Acisclo Boscan, 'A mis compatriotas. La independencia del Zulia', Baltimore (Maryland), June 1928.
197 AHMSGPRCP 1–10 Oct. 1928, Pérez Soto to Gómez, 10.10.28.
198 AHMSGPRCP 12–20 Junio, 1928, Argenis Azuaje to Gómez, 3.7.28.
199 AHMSGPRCP 1–10 Ago 1927, Leyba to Gómez, 9.8.27.
200 AHMSGPRCP 1–15 Sept. 1927, Leyba, 'Informe', 1.10.27; AHMSGPRCP 21–30 Nov. 1927, Leyba to Gómez, 28.11.27; and AHMSGPRCP 11–19 Dic. 1927, Azuaje to Gómez, 15.12.27.
201 AHMSGPRCP 11–20 Marzo, 1928, Pérez Soto, 'Memorándum No. 69 para el Benemérito General Juan Vicente Gómez', undated.
202 AHMCOP 277, Gómez to Azuaje, 12.6.28; and Ibid., Gómez to Leyba, 12.6.28.
203 Cf. AHMSGPRCS Nov.–Dic. 1928, *El Tiempo* (Bogotá), 'Pérez Soto trató de formar la República Zuliana arrebatándonos el Catatumbo', 14.10.28;

AHMSGPRCP 10–19 Ago 1928, Azuaje to Gómez, 19.7.28; and AHMSGPRCP 23–31 Dic. 1928, Rafael Esteban Párraga, 'Informe', 26.12.28.

204 AHMSGPRCP 21–31 Julio 1928, Pérez Soto and C. Salas to Gómez, 27.7.28. The small size of the armoury is surprising. It consisted of 202 rifles, distributed among the army in the State, and a further 113 in storage.

205 AHMSGPRCP 1–10 Oct. 1928, Leyba to Gómez, 7.10.28; and AHMSGPRCP 1–9 Julio 1931JVG, C.E. Pelayo to Gómez, 7.7.31.

206 AHMSGPRCP 1–10 Oct. 1928, Leyba to Gómez, 7.10.28.

207 AHMSGPRCP 21–31 Oct. 1928, Leyba to Gómez, 31.10.28.

208 AHMSGPRCP 11–20 Feb. 1929, Carlos B. Figueredo to Gómez, 18.2.29.

209 AHMSGPRCP 1–15 Sept. 1929, P. Itriago Chacín to J.M. Clemente, 24.9.29.

210 Ibid.

211 FO 371/13558, Donald Outon-Powell to Foreign Office, 11.6.29.

212 Rafael Simón Urbina, *Victoria, dolor y tragedia* (Caracas, Tip. Americana, 1936).

213 'Información de Maracaibo', *BCCC*, 17:177 (1.8.28), 4155–7.

214 Zulia, *Memoria y Cuenta, 1928*, Olivo, 'Resolución', 17.6.28, p. 127.

215 Ibid., M.A. Belloso, 'Relación de las sumas recaudadas para los damificados en el incendio de Lagunillas i su inversión', 15.9.28, pp. 100–1; and Ibid., Gómez to Pérez Soto, 22.6.28, p. 128.

216 MinRelInt, *Memoria 1928*, Pérez Soto to Interior Minister, 4.7.28, Doc. 36, pp. 83–5 (p. 83).

217 'Los Presidentes de Estados y los sucesos del año 28', *BAHM*, 2:7 (Julio–Ago 1960), 123–43, Pérez Soto to Gómez, 15.7.28.

218 MinRelInt, *Memoria 1928*, Pérez Soto to Interior Minister, 4.7.28, Doc. 36, pp. 83–5; Ibid., Arcaya to Pérez Soto, 7.7.28, p. 85; and Ibid., Pérez Soto to Arcaya, 14.7.28.

219 'Los Presidentes de estado y los sucesos del año 28', *BAHM*, Pérez Soto to Gómez, 15.7.28, p. 128.

220 Zulia, *Memoria y Cuenta, 1928*, Gonzalo Angarita *et al.* to Pérez Soto, 11.8.28, pp. 106–7.

221 Ibid., De Booy to Secretary-General of Zulia, 17.8.28, pp. 108–9.

222 AHMSGPRCP 21–31 Oct. 1928, M. Filiponel, 'Informe', 19.10.28.

223 AHMSGPRCP 21–31 Oct. 1928, Isaac Gómez to Gómez, 21.10.28.

224 AHMSGPRCP 20–22 Dic. 1928, José Jesús Paris to Alamo, 21.12.28. The same report went to Gómez.

225 Ibid.

226 Ibid.

227 Ibid.

228 Ibid.

229 The post was abolished in August 1931. (AHMSGPRCP 11–19 Ago 1931, Luis Nava G. to Requena, 11.8.31.)

230 MinRelInt, *Memoria 1929*, Rubén González to Alamo, 31.10.29, Doc. 222, pp. 219–21.

231 Luis F. Calvani, *Nuestro máximo problema* (Caracas, Editorial Grafolit,

1947), p. 49.
232 MinRelInt, *Memoria 1930*, González, 'Resolución', 3.10.30, Doc. 363, p. 384.
233 Ibid., Jonás González Miranda to González, 27.12.30, Doc. 365, pp. 385–92 (p. 387).
234 Ibid.
235 Ibid., Luis Nava R., 'Informe General que presenta al Ministerio de Relaciones Interiores el cuidadano Luis G. Nava R., Comisionado Especial ante las Compañías Petroleras para el cumplimiento de las disposiciones de la Ley de Trabajo en los Estados Falcón y Zulia', 31.12.30, Doc. 336, pp. 392–400 (p. 400).
236 Ibid., p. 394.
237 Ibid., p. 396.
238 Ibid., p. 400.
239 Zulia, *Memoria y Cuenta, 1930*, Asunción Soto *et al.* to Pedro L. Duno Pérez, 20.2.29, p. 70. They had a vested interest in this because of their joint ownership of a steamboat.
240 Ibid., Crebbs to Secretary-General, 2.3.29, p. 71.
241 AHMSGPRCP 1–15 Oct. 1929, Félix E. Durán *et al.* to Torres, 22.11.29.
242 Ibid.
243 AHMSGPRCS Julio 1930, Manuel Borjas H. to Gómez, 29.7.30. Gómez also donated land to the people of Encontrados to build new houses after a fire had destroyed the town. (AHMSGPRCP 20–30 Junio 1930JVG, Pérez Soto to Gómez, 24.6.30.)
244 AHMSGPRCP 11–20 Feb. 1931JVG, Solicitud al Inspector Técnico de Hidrocarburos, 22.1.31.
245 AHMSGPRCP 11–20 Feb. 1931JVG, Juan de Rios Vivas to Gómez, 19.2.31.
246 AHMSGPRCP 21–28 Feb. 1930, Bartolomé Osorio Quintero to Gómez, 28.2.30.
247 AHMSGPRCP 1–10 Junio 1933 *(sic)*, Pedro Paris to Gómez, 1.7.33.

5   Greater control of the oil industry

1 AHMSGPRCP 10–19 Enero 1924, Calvani to Gómez, 17.1.24.
2 AHMSGPRCP 17–31 Marzo 1925, Calvani to Alamo, 26.3.25.
3 MinFo, *Memoria 1924*, Exposición, p. xii.
4 FO 371/10653, Hobson to Foreign Office, 7.11.25.
5 AHMSGPRCP 1–16 Marzo 1925, Calvani to Gómez, 14.3.25.
6 Ibid.
7 MinFo, *Memoria 1924*, Doc. 24, pp. 40–1, For a full account see B.S. McBeth, *Royal Dutch-Shell vs. Venezuela* (Oxford Microfilm Publications, 1982), pp. 161–74.
8 DS 831.6363/290, Cook to Secretary of State, 28.9.25.
9 DS 831.6363/246, J. Webb Benton to Secretary of State, 4.10.24.
10 AHMSGPRCP 1–10 Enero 1927 *(sic)*, W.F. Buckley to Gómez, 24.1.28.
11 DS 831.6363/C73/4, Harry J. Anslinger to Secretary of State, 3.7.24.
12 The following companies were represented: Venezuelan Sun Oil Co.,

Gulf Oil, SOC(Cal), Exxon, British Controlled Oilfields Ltd, West India Oil Co. and Orinoco Oil Co. Shell declined the offer to attend. (DS 831.6363/C73/3, Chabot to Secretary of State, 5.7.24; and DS 831.6363/C73/1, Chabot to Secretary of State, 28.6.24.)

13 AHMSGPRCP 1–10 Enero 1927, *(sic)*, Buckley to Gómez, 17.7.24.
14 DS 831.6363/225, Chabot to Secretary of State, 12.6.24.
15 Ministerio de Obras Públicas, *Puerto petrolífero en Venezuela. Informe del Ingeniero del Gobierno sobre el establecimiento de un puerto de la Costa Occidental de la península de Paraguaná, Venezuela* (Madrid, Nuevas Gráficas, ND).
16 AHMSGPRCP 1–15 Mayo 1925, José Isidro Murillos to Gómez, 12.5.25.
17 AHMSGPRCP 1–10 Enero 1927 *(sic)*, Buckley to Gómez, 24.1.28.
18 Ibid. A little later it was known that Arcaya, owner of concessions on Paraguaná held by nominees, had transferred them to Exxon.
19 Thomas F. Lee, 'The rush for oil in Venezuela', *The World Today* (Dec. 1925), 355–68 (p. 363).
20 Ibid.
21 MinRelInt, *Memoria 1926*, Gómez, 'Decreto', 3.4.26, Doc. 8, pp. 11–12.
22 DS 831.6363/322, Sloan, 'The construction of a national port', 20.5.26.
23 AHMCOP 242, Baptista Galindo to M. Centeno Grau, 25.5.26.
24 DS 831.6363/366, Cook to Secretary of State, 3.10.27.
25 Ibid.
26 Ibid.
27 AHMSGPRCP 1–10 Enero 1927 *(sic)*, Buckley to Gómez, 24.1.28.
28 R. González and Lucio Baldó, 'Memorándum sobre la Compañía Venezolana de Petróleo, desde su fundación en junio de 1923 hasta el mes de octubre de 1920 *(sic)*', Caracas, 15.6.46, in Rafael González Rincones, *Cartas Barinesas* (Caracas, Editorial Sucre, 1958), pp. 212–4.
29 AHMSGPRCP 20–31 Dic. 1927, Antonio Andújar to Gómez, 27.12.27.
30 AHMSGPRCP 15–21 Sept. 1927, Carlos B. Figueredo to Gómez, 13.9.27.
31 AHMSGPRCP 1–9 Feb. 1928, Alamo to Sixto Tovar, 9.2.28.
32 AHMSGPRCP 1–9 Feb. 1928, 'España y Venezuela', *La Libertad* (Caracas), 9.2.28; and AHMSGPRCP 15–31 Ago 1932, J.M. Careaga Urquijo to Requena, 4.7.32.
33 AHMCOP 277, Gómez to Centeno Grau, 27.4.28.
34 DS 831.6363/390, Engert to Secretary of State, 20.6.28.
35 Ibid.
36 AHM, Martin Engineering Co., 'Proyecto para el puerto libre de Turiamo', Junio–Nov. 1928.
37 AGTCOP 14, Torres to Betancourt Sucre, 12.1.30.
38 Ibid.
39 MinFo, *Memoria 1927*, Siro Vásquez *et al.*, 'Informe de la Comisión nombrada para estudiar los derrames de petróleo en las aguas del Lago de Maracaibo', 5.2.28, pp. 409–18 (p. 409).
40 DS 831.6363/395, Sloan, 'The contamination of the waters of Lake Maracaibo', 9.7.28.
41 Ibid.
42 DS 831.6363/396, Cook to Secretary of State, 20.7.28.
43 MinFo, *Memoria 1928*, Exposición, p. xvii.

44 AHMSGPRCP 1–10 Marzo 1928, Alamo to Gómez, 8.3.28.
45 MinRelInt, *Memoria 1928,* p. 309–10.
46 The Royal bank of Canada, *Business conditions in Latin America and the West Indies* (Montreal) Sept. 1928, p. 13.
47 'Situación petrolera', *BCCC,* 18:186 (1.5.29), 4390.
48 MinFo, *Memoria 1929,* Exposición, p. vii.
49 AHMSGPRCP 1–15 Sept. 1929, Calvani to Gómez, 1.9.29.
50 MinFo, *Memoria 1929,* p. 346.
51 Luis Mariñas Otero, *Las constituciones de Venezuela* (Madrid, Ediciones Cultura Hispánica, 1965), p. 81. Quoting from article 128 of the Constitution.
52 MinFo, *Memoria 1929,* Exposición, p. xxiv.
53 MinFo, *Memoria 1930,* Exposición, p. xx.
54 MinFo, *Memoria 1929,* Exposición, p. xxviii.
55 DS 831.6363/433, Engert to Secretary of State, 25.11.29.
56 Cf. Francisco J. Parra, *Historia de las leyes. Ley sobre liquidación de derechos de boyas en la aduana de Maracaibo* (Caracas, Tip. Americana, 1943).
57 DS 831.6363/Lago Pet./L4, Engert to Secretary of State, 22.12.29. Enclosure, W.T.S. Doyle, 'Memorandum on the Buoy Tax Situation', 10.12.29.
58 Doyle, 'Memorandum', 10.12.29.
59 Ibid.
60 AHMSGPRCP 1–15 Julio 1929, 'Memorándum presentado por Lago Petroleum Corporation, The Caribbean Petroleum Company, The Colon Development Company Limited, and the Venezuelan Oil Concessions Limited', 12.7.29.
61 AHMSGPRCP 1–14 Ago 1929, D. Giménez to Gómez, 9.8.29.
62 Doyle, 'Memorandum', 12.10.29.
63 AGTAPM 1928 *(sic),* Torres, 'Historia de una Comisión del servicio público', 26.10.29.
64 Ibid.
65 Interview Dr P.A. Capiola Torres, 27.9.77.
66 AGTAPM 1928 *(sic),* Torres, 'Historia', 26.10.29.
67 DS 831.6363/Lago Pet./L4, Engert to Secretary of State, 22.12.29.
68 Ibid., Enclosure 2 Doyle to Engert, 17.12.29.
69 Ibid.
70 Ibid.
71 Ibid.
72 Ibid.
73 DS 831.6363/Lago Pet./15, Engert to Secretary of State, 20.2.30.
74 *Gaceta Oficial,* No. 17,055, 27.2.30.
75 MinFo, *Memoria 1930,* Exposición, p. xxiii.
76 AHMSGPRCS 16–30 Feb. 1932, 'Memorándum', undated and unsigned.
77 Rodolfo G. Rincón, *Contra una injusticia* (Caracas, Editorial Bolívar, 1936).
78 Cf. Manuel Matos Romero, *El problema petrolero en Venezuela* (Caracas, Editorial Bolívar, 1938); and Betancourt, *Venezuela. Política y petróleo,* 2nd edn (Caracas, Editorial Senderos, 1967).
79 AHMSGPRCS 16–30 Feb. 1932, 'Memorándum', undated and unsigned. Another method employed by the companies to obtain the buoy tax

reduction was to convince the Customs that the oil tankers carried agricultural exports which in fact consisted only of a few goats and bananas on their decks. (Ibid.).

80 AGTC Ago–Dic. 1930, Draft Torres to Doyle, F.C. Pannill and José Santiago Rodríguez, Oct. 1930; and, AGTAPM 1930, 'Memorándum: Asunto Boyas', undated and unsigned.
81 AGTC Enero–Junio 1931, Torres to Villegas Pulido, 2.1.31.
82 AHMV and AGTC Enero–Junio 1931, Torres to Leyba, 26.1.31. This was to be done for 15–20 tankers.
83 APACT, Torres to Pedro Antonio, 14.9.42.
84 AGTC Julio–Dic. 1931, Salvador T. Maldonado, 'Apuntes especiales sobre asuntos relacionados con el petróleo y sus impuestos, dedicados al señor Doctor Gumersindo Torres, Ministro de Fomento', 2.2.31.
85 AGTC Enero–Julio 1930, Torres to Alberto Urbaneja, 18.6.30.
86 AGTC Enero–Julio 1930, Torres to Clemente, 24.1.30.
87 'Guillermo Zuloaga 1930–1970. Las Inspectorías Técnica de Hidrocarburos', *El Farol* (Caracas), Suplemento al No. 234, undated.
88 MinFo, *Memoria 1930*, Exposición, p. xvii.
89 Ibid., Exposición pp. xvii–xviii. For example, C.A. Velutini, Field Inspector, had worked previously for Texaco and Sun Oil Co.
90 This idea was first mooted in 1916 by Díaz Rodríguez, then the Minister of Development.
91 MinFo, *Memoria 1930*, Exposición, p. xviii.
92 MinFo, *Memoria 1931*, p. 299.
93 AHMSGPRCP 1–20 Abril 1932, R. Alvárez de Lugo, 'Observaciones sobre la carta anónima de fecha 1 de febrero último dirigida al sr. Dr. Rafael Requena, relativa a las compañías petroleras establecidas en el país', 15.4.32.
94 FO 371/14300, O'Reilly to A. Henderson, 11.9.30.
95 AGTC Ago–Dic. 1930, Leon C. Booker to Torres, 19.9.30.
96 AGTC Ago–Dic. 1930, Torres to Booker, 20.9.30.
97 Gumersindo Torres, *Memorándum de algunas Compañías petroleras sobre el Reglamento de la Ley de Hidrocarburos y demás Minerales Combustibles y Observaciones que el Ministro de Fomento hace a dicho Memorándum* (Caracas, Tip. Central, 1930), p. 19.
98 AGTC Ago–Dic 1930, Torres to Zuloaga, 20.9.30.
99 FO 371/14300, O'Reilly to Henderson, 11.9.30.
100 Ibid.
101 AGTC Ago–Dic. 1930, Torres to Zuloaga, 20.9.30.
102 Cf. Torres, *Memorándum de algunas compañías*.
103 AGTCOP 15, Torres to Luis Gerónimo Pietri, 16.10.30.
104 Torres, *Memorándum de algunas compañías*, p. 19.
105 Ibid., pp. 20–1.
106 Ibid.
107 Cf. Alejandro Pietri, *Lago Petroleum Corporation, Standard Oil Company of Venezuela y Compañía de Petróleo Lago contra la Nación por la negativa de exoneración de derechos de importación* (Caracas, Lit. y Tip. del Comercio, 1940), p. 324.
108 Torres, *Memorándum de algunas compañías*, p. 39.

109 Ibid., p. 24.
110 Ibid., p. 28.
111 Ibid., p. 28.
112 Ibid., p. 30.
113 Ibid., p. 31.
114 Ibid., p. 31.
115 Ibid., p. 31.
116 Ibid., p. 33.
117 Ibid., p. 34.
118 Ibid., p. 37.
119 Ibid., p. 38.
120 AMF, Torres to Zuloaga, 6.10.30.
121 AMF, Torres to Zuloaga, 11.10.30. This tight control of production would be of great benefit to the local concessionaires, including the government, who retained production royalties with the companies.
122 MinFo, *Memoria 1930,* Exposición, p. xxi.
123 AGTAPM 1929 *(sic),* and FO 199/268, Torres to Doyle, E. Aguerrevere, and Booker, 18.10.30.
124 Ibid.
125 AGTC Ago–Dic. 1930, Velutini to Torres, 9.10.30.
126 FO 371/17619, Keeling to Foreign Office, 1.2.34, Enclosure, R. Cayama Martínez to Oil Companies, 22.8.31.
127 The companies on average deducted 67 cents from the price of oil, whereas the average transport cost per barrel was 23 cents.
128 AHMSGPRCPJBPérez, 1–30 Abril 1931, Torres to Jordan Herbert Stabler, 14.10.30, in Stabler to Torres, 24.4.31.
129 AGTC Enero–Junio 1931, Torres to Calvani, 24.2.31.
130 AHMSGPRCPJBPérez 1–30 Abril 1931, Stabler to Torres, 24.4.31.
131 Ibid.
132 Ibid.
133 AGTC Enero–Junio 1931, and AHMSGPRCPJBPérez 1–30 Abril 1931, Torres to Stabler, 25.4.31.
134 Ibid.
135 Ibid.
136 Ibid.
137 Ibid.
138 AHMSGPRCPJBPérez 1–31 Mayo 1931, Stabler to Torres, 2.5.31.
139 Ibid.
140 AHMSGPRCPJBPérez 1–31 Mayo 1931, Stabler to J.B. Pérez, 4.5.31.
141 AGTCOP 16, Torres to Pedro (Arcaya), 24.4.31.
142 AGTC Enero–Junio 1931, Arcaya to Torres, 26.5.31.
143 'Circular dirigida por el Ministro de Fomento a las Compañías de Petróleo, acerca de algunas dificultades encontradas para la inspección y fiscalización oficial, Caracas, 30.5.31', *Boletín del Ministerio de Fomento* (Tercera Epoca), 1:11 (1.7.31.), 1084–5 (p. 1085).
144 AGTC Enero–Junio 1931, E. López Contreras to Torres, 16.6.31.
145 AHMSGPRCP 1–9 Sept. 1931 *(sic)*, Ministro de Fomento to Stabler, undated.
146 AGTC Julio–Dic. 1931, 'Memorándum', undated and unsigned.

147 AGTC Enero–Junio 1931, Torres to Gómez, 18.6.31.
148 AGTC Julio–Dic. 1931, 'Memorándum', 4.7.31.
149 Ibid.
150 AHMSGPRCS 15–30 Sept. 1931, Stabler to Requena, 26.9.31.
151 AHMSGPRCP 1–10 Oct. 1931, 'Memorándum sobre contestación del Ministro de la nueva proposición de la 'Venezuela Gulf Oil Company' acerca de la base para fijar el precio del petróleo y liquidar el impuesto de explotación correspondiente', 8.10.31.
152 AHMSGPRCP 11–20 Sept. 1933, Luis Herrera F., 'Informe que al Ministro de Fomento presenta la Inspectoría Técnica General de Hidrocarburos referente al Memorándum que el Dr. R. González Miranda dirigió al Ciudadano Presidente de la República respecto a la manera de obtener mejores provechos en los impuestos sobre el petróleo', 21.9.33.
153 This difference was considerably enhanced when the U.S. imposed in June 1932 a tariff on imported oil. The value of Venezuelan oil then dropped to $0.32 ($0.53 minus the tariff, $0.21), making the value of a ton of oil $2.02 or Bs.10.5. The Government's 10 per cent royalty would then be Bs.0.95 below the minimum established by law, and Bs.0.71 in the case of Lago. (Ibid.)
154 AHMSGPRCS 1–15 Nov. 1931, Cayama Martínez to Requena, 4.11.31. Another claim against Gulf Oil was that several of its concessions illegally paid production royalties at 11 per cent instead of 15 per cent. Salvador T. Maldonado, President of the Contaduría General of the Finance Ministry, consulted Calvani over this matter because he was thinking of bringing a claim against the company. Calvani counselled that he should wait since the time was not propitious for such an action. Several years later, after Gómez's death, on 28 July 1937, the Attorney-General brought the suit against the company claiming Bs.15,630,200 in back taxes and Bs.11,000,298.2 in interest accruing on the first sum. The Federal Court of Cassation found the company guilty and ordered it to pay the Government Bs.15,625,491.98. (Cf. Luis F. Calvani, *Nuestro máximo problema* (Caracas, Editorial Grafolit, 1947), p. 30; and Corte Federal y de Casación, *Memoria 1938*, Sentencia 12.)
155 AMF, Torres to Zuloaga, 31.1.31.
156 F.G. Rapport to Torres, 23.3.31, in *Boletín del Ministerio de Fomento* (Tercera Epoca), 1:8 (15.4.31), 777–81.
157 Ibid., Torres to Rapport, 30.3.31, 781–5.
158 AGTC Enero–Junio 1931, Draft Torres to Doyle, Feb. 1931.
159 MinFo, *Memoria 1930*, Exposición, p. il.
160 AGTC Enero–Julio 1930, Torres to Clemente, 24.1.30.
161 MinFo, *Memoria 1930*, Exposición, p. xlviii.
162 'Circular dirigida a los Representantes de The Carribbean Petroleum Company, Lago Petroleum Corporation and West India Oil Company', *Boletín del Ministerio de Fomento* (Tercera Epoca), 1:6 (22.1.31.), 548–9 (p. 549).
163 MinFo, *Memoria 1930*, Exposición.
164 AHMSGPRCP 21–31 Julio 1928, Aquiles Iturbe, 'Memorial para el general Juan Vicente Gómez', 30.7.28.
165 AHMSGPRCP 16–30 Nov. 1932 *(sic)*, 'Memorándum sobre la con-

veniencia de establecer en el país una refinería nacional para el petróleo crudo proveniente de la participación ó royalty que recibe el gobierno de numerosas concesiones de explotación', 26.11.29.

166 Calvani, *Nuestro máximo problema*, pp. 28–9.
167 AGTC Enero–Junio 1931, Draft Torres to Zuloaga, undated.
168 AGTC Ago–Dic. 1930, Torres to P.R. Rincones, 27.10.30.
169 AHMSGPRCP 21–31 Sept. 1930JVG, P.S. Luigi to Gómez, 28.9.30.
170 AHMSGPRCP 1–30 Nov. 1930JBPérez, Luigi, 'Proyecto de contrato', undated.
171 AHMSGPRCS Nov.–Dic. 1930, Draft, 'Contrato que celebra el ciudadano Ministro de Fomento . . . y el ciudadano Dr. Pedro Simón Luigi', undated.
172 AHMSGPRCP 1–10 Nov. 1930JVG, Luigi to Gómez, 6.11.30.
173 AGTC Ago–Dic. 1930, Torres, 'Memorándum', 28.10.30.
174 Ibid.
175 AGTC Ago–Dic. 1930, Zuloaga to Torres, 13.11.30.
176 Ibid.
177 MinFo, *Memoria 1914*, Exposición, pp. xi–xii.
178 AGTC Ago–Dic. 1930, Raymond Concrete Pile Co. to Torres, 8.12.30.
179 AGTC Enero–Junio 1931, Torres to Rincones, 14.4.31.
180 AHMSGPRCP 1–15 Dic. 1931, L.C. Chase to Gómez, 1.12.31.
181 AHMSGPRCS 11–19 Junio 1932, McGoodwin to Requena, 18.6.32.
182 Ibid. The price was certainly higher than the average paid by Exxon and Lago, which between 1929 and 1932 was Bs.16.93 per barrel.
183 AHMSGPRCS 16–31 Marzo 1932, Draft, 'Constitución de la Compañía Anónima Nacional de Industrias Petroleras', undated.
184 AHMSGPRCS 16–31 Marzo 1932, P.R. Tinoco, 'Memorándum para el Benemérito General Juan Vicente Gómez', 21.3.32.
185 Ibid.
186 AHMSGPRCP 16–31 Enero 1933, J.W. Stewart to Requena, 21.1.33.
187 AHMSGPRCP 1–28 Feb. 1931JBPérez, Arcaya to Gómez, 31.3.31.
188 AHMSGPRCP 1–14 Marzo 1931JVG, Translation, *Journal of Commerce*, 4.3.31.
189 U.S. House of Representatives, 'Production costs of crude petroleum and refined petroleum products', *House Document No. 195*, 72 cong. 1 Sess., 1932, Table 25, p. 49.
190 AHMSGPRCP 21–29 Feb 1932, E.A. van Cleck to Senators Robert F. Wagner and Royal S. Copeland, 29.2.32.
191 U.S. Tariff Commission, *Petroleum* (War Changes in Industry Series No. 17, 1946), p. 75.
192 Cf. J.D. Butler, 'The influences of economic factors on the location of oil refineries (with primary reference to the world outside the USA and USSR)', *The Journal of Industrial Economics* 1:3 (July 1959), 187–201, Table 2, p. 190. In addition, Venezuela became the most important source of oil for the UK in the event of war and the US being 'unfriendly'. (CAB 50/3/Secret/O.B. 27, Committee of Imperial Defence, Oil Board, Sub-committee, 'Report on oil supply in time of war', 20 March 1929.)
193 U.S. House of Representatives, *Petroleum Investigation*, Hearings before a

subcommittee of the Committee of Interstate and Foreign Commerce, Part 1, 73 Cong. (Recess), 17–22 Sept. 1934, Statement, Arthur H. Redfield.
194 U.S. Tariff Commission, *Petroleum.*
195 Corte Federal y de Casación, *Memoria 1932*, Sentencia 13, pp. 180–97.
196 Ibid., Docs. 19 and 25, pp. 211–36 and 248–65 respectively.
197 AHMSGPRCP 20–31 Julio, Rufino González Miranda, 'Memorándum', undated.
198 Ibid., González assumed that Venezuela produced predominantly high grade crude, which was not the case.
199 Ibid.
200 AHMSGPRCP 11–20 Sept. 1933, Luis Herrera F., 'Informe', Caracas, 21.9.33. Herrera based his calculation on the assumption that 500 additional inspectors would be needed. At a daily salary of Bs.20 the total wage bill for a 360-day year would amount to Bs.3,600,000.
201 Ibid.
202 Ibid.
203 Ibid.
204 MinFo, *Memoria 1933*, Exposición.
205 Ibid.
206 MinFo, *Memoria 1934*, Exposición, p. xxi.
207 MinFo, *Memoria 1935*, Doc. 84, Juan B. Carroz *et al.* to *Jefe Civil* del Distrito Urdaneta, 3.6.35; and Ibid., Doc. 85, Tinoco to J.M. Leonardi Villasmil, 12.8.35.
208 AHMSGPRCP 1–15 Sept. 1935, Arcaya to Gómez, 11.9.35.
209 AHMSGPRCP 1–15 Oct. 1935, Arcaya to Gómez, 9.10.35.
210 Ibid.
211 Ibid.
212 MinFo, *Memoria 1935*, Exposición, pp. xiv–xv.
213 FO 371/19846, Keeling to Harvey, 24.7.36.
214 FO 371/19846, F. Godber to F.C. Starling, 11.9.36.
215 Ibid.
216 The companies acted together previously during 1921–2, but this was the result of the U.S. Legation's initiative.

Bibliography

ARCHIVES

United Kingdom

Public Record Office
Cabinet Office
CAB 50 Committee of Imperial Defence, Oil Board, 1925–39
Foreign Office
FO 80 General Correspondence, 1900–5
FO 368 General Correspondence, Commercial, 1906–19
FO 369 General Correspondence, Consular, 1906–33
FO 370 General Correspondence, Library, 1906–33
FO 371 General Correspondence, Political, 1906–36
FO 420 General Correspondence, Confidential Print, America, South and
Central, 1908–36
FO 115 U.S. Embassy and Consular Archives, 1908–29
FO 199 Venezuela. Embassy and Consular Archives, 1904–33

Science Museum Library (London)
S. Pearson & Sons Ltd Archives
Box A 5–7
Box C-25, C-30

United States of America

Records of the Department of State relating to the Internal Affairs of
Venezuela, 1910–29. National Film Archives Microcopy No. 366, Reels 24–8

Venezuela

Fundación John Boulton
Archivo del General Antonio Aranguren, Files 1–8

Ministerio de Energía y Minas
Archivo dependiente de la División de Conservación de la Oficina Técnica de
  Hidrocarburos
Traspasos, 1925–35
Historial de Concesiones de Hidrocarburos, Vols. 1–35
Archivo del Ministerio de Fomento. Uncatalogued
Archivo del Ministerio de Fomento, Hidrocarburos. Correspondencia,
  uncatalogued
Inspectoría Técnica de Minas – Informes sobre hidrocarburos, 1921–35

Palacio de Miraflores
Archivo Histórico de Miraflores
Presidential copybooks, Nos. 100–362, 1908–36
Presidential and Secretary-General's correspondence, 985 file bundles,
  1908–36

Private archives
Archivo Particular del Dr Gumersindo Torres
Copybooks, 1917–22, 1928–31
Correspondence, 1917–22, 1928–31
Archivo del Dr J.M. Capiola Torres
Various letters

CONTEMPORARY OFFICIAL PUBLICATIONS

United Kingdom

Board of Trade Journal and Commercial Gazette, 1908–36

United States of America

Congressional Record. Proceedings and Debates, 1900–35
Department of State, *Papers relating to the Foreign Relations of the United States of
  America–Venezuela,* 1900–36
Department of Commerce and Labor, Bureau of Manufacture, *Monthly Consular
  and Trade Reports – Venezuela,* 1900–36

Venezuela

Banco Agrícola y Pecuario, *Informe,* 1931
Banco Central de Venezuela, *Informe,* 1945
Consejo de Gobierno, *Memoria,* 1909–14
Consulate General of Venezuela (Liverpool), *The Venezuelan Commercial Review,*
  1933–5
Corte Federal y de Casación, *Memoria,* 1909–36
Diario de Debates de la Cámera de Disputados, 1918–36
Diario de Debates de la Cámara del Senado, 1909–19, 1922–36

Distrito Federal, Gobernación, *Exposición*, 1911–35
Gaceta Oficial, 1909–10, 1925
Ministerio de Fomento, *Anuario Estadístico de Venezuela*, 1908–12, 1938
  Cuenta, 1913–36
  Memoria, 1908–36
  Dirección General de Estadística, Estadística mercantil y marítima, 1907–30
Ministerio de Hacienda, *Memoria*, 1908–36
Ministerio de Obras Públicas, *Memoria*, 1908–36
Ministerio de Relaciones Exteriores, *Libro Amarillo*, 1908–36
  Cuenta, 1908–36
Ministerio de Relaciones Interiores, *Memoria*, 1908–36
Recopilación de Leyes y Decretos de Venezuela, Vols. 1–60
Zulia (Estado), *Memoria y Cuenta* 1908, 1911, 1916–19, 1922–35
  Presidencia, Mensaje, 1911–12, 1916–18, 1921–4, 1929, 1931, 1935

OTHER CONTEMPORARY SOURCES

El Agricultor Venezolano (Ministerio de Agricultura y Cría), Caracas, Nos. 31–48, 1938–42
Boletín de la Cámara de Comercio de Caracas, Caracas, 1919–36
Boletín del Archivo Histórico de Miraflores, Caracas, Nos. 1–100, 1959–77
Boletín del Archivo Nacional, Caracas, 1964–71.
Boletín Comercial e Industrial (Ministerio de Relaciones Exteriores), Caracas, 1920–4
Boletín del Ministerio de Fomento, Caracas, 1909–35
Boletín del Ministerio de Relaciones Exteriores, Caracas, Vols. 1–10, 1908–39
Boletín del Ministerio de Relaciones Exteriores, Número Extraordinario, Caracas, Vols. 1, 3, 11, 1925–32
Boletín del Ministerio de Relaciones Exteriores, Supplemento Comercial é Industrial, Caracas, Nos. 1–4, 1930
Boletín del Petróleo, Caracas, Nos. 1–37, Abril–Dic. 1925
Corporation of Foreign Bond-Holders (London), Newspaper Files, Vols. 7–11
El Eco Alemán, Caracas, 1916
Latin American World, London, 1925–32
Oil Facts and Figures, London, 1912–36
Oil News, London, 1913–36
Petroleum Times, London, 1919–36
Petroleum World, London, 1915–31
The Pipeline, London, Vols. 1–14, 1920–34
Revista del Instituto Nacional de Café, Caracas, Vols. 1–3, 1940–3
Royal Bank of Canada, *Business Conditions in Latin America and the West Indies*, Montreal, 1928–36
Royal Dutch Company, *Annual Reports*, London, 1907–36
The 'Shell' Transport and Trading Company, *Annual Reports*, London, 1907–36
Walter R. Skinner, *The Oil and Petroleum Manual*, London, 1910–36
The South American Journal, London, 1900–36
The Stock Exchange Yearbook, London, 1900–14
Venezuela Contemporánea, Caracas, 1916–17

Venezuela, Comercial, Social é Intelectual, Caracas, 1924–7
World Petroleum, London, 1930–1

NEWSPAPERS

El Constitucional, Caracas, 1908–9
La Información, Maracaibo, 1921–2, 1926
El Nuevo Diario, Caracas, 1913, 1918–22, 1928–32
Panorama, Maracaibo, 1920–23, 1935
The Times, London, 1908–9
El Universal, Caracas, 1918–22, 1928–32

UNPUBLISHED SOURCES

Carlisle, Douglas H., 'The organization for the conduct of foreign relations in Venezuela, 1909–1935', Ph.D.Diss., University of North Carolina at Chapel Hill, 1951
Knudson, David Lawrence Taylor, 'Petroleum, Venezuela, and the United States: 1920–1941', Ph.D.Diss., Michigan State University, 1975
McBeth, B.S., 'Juan Vicente Gómez and the Venezuelan oil industry; with special relevance to the British oil companies', B.Phil.Diss., Oxford University, 1975
'British oil supplies, 1919–1939', 1978 Ms
'Juan Vicente Gómez and the oil companies', D.Phil.Diss., Oxford University, 1980
Moore, John Robert, 'The impact of foreign direct investment on an underdeveloped economy. The Venezuelan case', Ph.D.Diss., Cornell University, 1956
Redfield, A.H., 'Our petroleum diplomacy in Latin America', Ph.D.Diss., The American University, 1942
Stann, E. Jeffrey, 'Caracas, Venezuela 1891–1936: a study of urban growth', Ph.D.Diss., Vanderbilt University, 1975

INTERVIEWS

Dr Antonio Aranguren Fonseca, Caracas, 14.9.77
Dr P.A. Capiola Torres, Caracas, 27.9.77 and 29.9.77
Sra Carmen Carolina Torres de Lecuna, Caracas, 27.9.77
Dr Manuel R. Egaña, Caracas, 13.10.76
Dr César González, Caracas, 23.9.76 and 16.1.77
Dr J.A. Giacopini Zárraga, Caracas, 1.4.77 and 25.7.77
Sr Gustavo Machado, Caracas, 1.9.76
Dr Pedro José Muñoz, Caracas, 7.9.76
Dr Juan Pablo Pérez Alfonzo, Caracas, 19.4.77, 21.5.77 and 29.8.77
Dr Amenodoro Rangel Lamus, Caracas, 2.7.77
Senator Dr Ramón J. Velásquez, Caracas, 9.1.77, 5.2.77, 20.3.77, 9.7.77 and 21.8.77.
Dr Guillermo Zuloaga, Caracas, 28.9.77

OFFICIAL PUBLICATIONS

United Kingdom

Board of Trade, Department of Overseas Trade
Bancroft-Livingston, H., *Report of the economic and financial conditions in Venezuela*, London, HMSO 1927

Diplomatic and Consular Reports
Ifor Rees, T., Vice-Consul, 'Venezuela – Report for the year 1913–14 on the trade of Venezuela and the Consular District of Caracas', *PP*, Vol. LXXV, 1915, 849–81
Gilliat-Smith, Guy, Vice-Consul, 'Diplomatic and Consular Reports. Venezuela: Report for the year 1909–10 on the trade of Venezuela and the Consular District of Caracas', *PP*, 1911, XLVII, 767–807

United States of America

Department of Commerce and Labor, Bureau of Manufactures
Investment in Venezuela, Washington, USGPO, 1953

Department of Trade, Bureau of Foreign and Domestic Commerce
Bell, P.L., 'Venezuela; a commercial and industrial handbook', *Special Agents Series*, No. 212, 1922
Brett, H., 'Venezuela', *Supplement to Commerce Reports*, No. 51A, 1920
Dean, Charles J., 'Commerce and industrial development in Venezuela', *Trade Information Bulletin*, No. 783, 1931
Fletcher, S.J., 'Venezuela', *Supplement to Commerce Reports*, No. 13, 1922

Federal Trade Commission
Economic report on the investigation of coffee prices. Summary and conclusions, Washing- USGPO, 1954

House of Representatives
'Production costs of crude petroleum and refined products', Letter from the Chairman of the United States Tariff Commission. Report of an investigation made by the United States Tariff Commission relative to the cost of production of crude petroleum, fuel oil, gasoline and lubricating oils, produced in the United States and in specified foreign countries, *House Document No. 195*, 72 Cong. 1 Sess., 1932
Petroleum Investigation, Parts 1–4, Hearings before a subcommittee of the Committee on Interstate and Foreign Commerce, 73 Cong. (Recess), 1934

Senate
'Diplomatic correspondence with Colombia in connection with the Treaty of 1914 and certain oil concessions', *Senate Document No. 64*, 68 Cong. 1 Sess., 1924

'Cost of crude petroleum in 1931', U.S. Tarif Commission, *Senate Document No. 267*, 71 Cong. 3 Sess., 1931

American petroleum interests in foreign countries, Hearings before a Special Committee Investigating Petroleum Resources, 79 Cong. 1 Sess., 1946

Tariff Commission

The foreign trade of Latin America. A report on the trade with Latin America with a special reference to trade with the United States under the provision of title III, part II, section 332 of the Tariff Act of 1930, Washington, USGPO, 1942

Petroleum, War Changes in Industry Series No. 17, 1946

Venezuela

El General Juan Vicente Gómez. Documentos para la historia de su Gobierno (compiled by Luis Correa), Caracas, Lit. del Comercio, 1925

Banco Central de Venezuela

Ingreso Nacional de Venezuela en 1936, Caracas, Monografías del Banco Central de Venezuela No. 1, Editorial Relámpago, 1949

Jurado de Responsabilidad Civil y Administrativa

Sentencias, 5 vols., Caracas, Imprenta Nacional, 1946

Ministerio de Fomento, Dirección General de Estadística

Números índices de precios al por mayor (1913–1937), Caracas, Lit. y Tip. Casa de Especialidades, 1937

Resumén general de población del sexto censo nacional, Caracas, Tip. Garrido, 1938

Estadística del petróleo para los años 1936, 1937, 1938, Caracas, 1940

Ministerio de Fomento, Dirección de Estadística, Estado Zulia

Censo industrial, comercial y empresas que prestan servicios – 1936, Caracas, Tip. Garrido, 1939

Ministerio de Minas e Hidrocarburos, Dirección General, Division de Economía Petrolera

Petróleo y otros datos estadísticos, Caracas, 1964

Venezuelan petroleum industry. Statistical data, Caracas, Ministry of Mines and Hydrocarbons, 1966

Ministerio de Obras Públicas

Puerto petrolífero en Venezuela. Informe del Ingeniero del Gobierno sobre el establecimiento de la península de Paraguaná, Venezuela, Madrid, Nuevas Gráficas, ND

SECONDARY SOURCES

Books

Adriani, Alberto, *Labor venezolanista*, Caracas, Tip. La Nación, 1937

Bibliography 265

Arcaya, Pedro Manuel, *Memorias del Doctor Pedro Manuel Arcaya*, Madrid, Talleres del Instituto Geográfico y Catastral, 1963

Arnold, Ralph, MacGready, George A. and Barrington, Thomas W., *The first big oil hunt: Venezuela 1911–1936*, New York, Vantage Press, 1960

Betancourt, Rómulo, *Venezuela. Política y petróleo*, 2nd edn., Caracas, Editorial Senderos, 1967

Briceño Parilli, A.J., *Las migraciones internas y los municipios petroleros*, Caracas, Tip. ABC, 1947.

Brito Figueroa, Federico, *Venezuela siglo XX*, La Habana, Casa de las Américas, 1967

Calvani, Luis F., *Nuestro máximo problema*, Caracas, Editorial Grafolit, 1947

Compañía Venezolana de Petróleo, *Estatutos de la Compañía Anónima 'Compañía Venezolana de Petróleo'*, Caracas, 1923

Davenport, E.H. and Cooke,S.R., *The oil trusts and Anglo-American relations*, London, Macmillan and Co., 1923

Delgado Palacios, G., *Contribución al estudio del café en Venezuela*, Caracas, Publicaciones de la Junta Central de Aclimitación y Perfeccionamiento Industrial, Tip. El Cojo, 1895

De Sucre, María de Lourdes Acedo and Nones Mendoza, Carmen Margarita, *La Generación Venzolana del 1928. Estudio de una élite politica*, Caracas, Editorial Ariel, 1976

Deterding, Sir Henri, *An international oilman*, London, Ivor Nicholson and Watson Ltd, 1934

Egaña, Manuel R., *Tres décadas de producción petrolera*, Caracas, Tip. Americana, 1947

Gibb, George S. and Knowlton, E.H., *The resurgent years, 1911–1927*, New York, Harper and Bros., 1956

Gómez, J.V., *Mensaje que el ciudadano General J.V. Gómez, Presidente de los EE.UU. de Venezuela, presenta al Congreso Nacional en sus sesiones ordinarias de 1935*, Caracas, Lit. del Comercio, 1935

González Miranda, Rufino, *Estudios acerca del régimen legal del petróleo en Venezuela*, Caracas, UCV, 1958

González Rincones, Rafael (compiler), *Pioneros del petróleo en Venezuela*, Caracas, Editorial Sucre, 1956

Cartas barinesas, Caracas, Editorial Sucre, 1958

Guardia, Hugo N., *Recopilación de Estadística cafetera*, Caracas, Publicaciones del Instituto Nacional del Café, 1943

Henao Jaramillo, Jaime, *La caficultura y la economía nacional*, Caracas, Cuadernos Verdes No. 84, Tip. El Compás, 1950

Issawi, Charles and Yeganeh, Mohammed, *The economics of Middle Eastern Oil*, London, Faber and Faber, 1963

Lecuna, Vicente, *El historiador Vicente Lecuna y nuestra riqueza petrolera*, Caracas, Fundación Eugenio Mendoza, 1975

Lieuwen, Edwin, *Petroleum in Venezuela*, Berkeley, University of California Press, 1954

Linares, José, Antonio, *El General Juan Vicente Gómez y las obras públicas en Venezuela, 19 de diciembre de 1908–4 de agosto de 1913*, Caracas, Lit. y Tip. del Comercio, 1916

Luzardo, Rodolfo, *Venezuela. Business and finances*, Englewood Cliffs, NJ, Prentice-Hall, 1957

McBeth, B.S. *Royal Dutch-Shell vs. Venezuela*, Oxford Microfilm Publications, 1982

Mariñas Otero, Luis, *Las constituciones de Venezuela*, Madrid, Ediciones Cultura Hispánica, 1965

Matos Romero, Manuel, *El problema petrolero en Venezuela*, Caracas, Editorial Bolívar, 1938

Mendoza, José Rafael, *Juicio propuesto por el Concejo Municipal del Distrito Bolívar del Estado Zulia contra The Venezuelan Oil Concessions Limited por nulidad del contrato de 13 de octubre de 1926 sobre ejidos en Lagunillas, Cabimas, y demás Municipios del Distrito Bolívar. (Demanda, contrato, informes, sentencias de Primera Instancia)*, Caracas, Impresores Unidos, 1945

Meyer, Lorenzo, *México y Estados Unidos en el conflicto petrolero, 1917–1942*, México, El Colegio de México, 1968

Nass, Hermán, *El crédito agrícola en Venezuela*, Caracas, Cuadernos Verdes No. 14, Editorial Crisol, 1945

Parra, Alirio, *La industria petrolera y sus obligaciones fiscales en Venezuela*, Caracas, Primer Congreso Venezolano del Petróleo, 1962

Parra, Francisco J., *Historia de las leyes. Ley sobre liquidación de derechos de boyas en la aduana de Maracaibo*, Caracas, Tip. Americana, 1943

Pietri, Alejandro, *Lago Petroleum Corporation, Standard Oil Company of Venezuela y Compañía de Petróleo Lago contra la Nación por la negativa de exoner   ación de derechos de importación*, Caracas, Lit. y Tip. del Comercio, 1940

Prieto Soto, Jesús, *Huellas históricas de Cabimas*, México, Editorial Cultura, 1959

Quintero, Rodolfo, *Antropología de las ciudades latinoamericanas*, Caracas, UCV, 1965

La cultura del petróleo, Caracas, Colección Esquema Gráfica Universitaria C.A., 1968

Rangel, Domingo Alberto, *Los Andinos en el poder*, Caracas, 1964

Capital y Desarrollo. El Rey Petróleo, 2 vols., Caracas, UCV, 1970

Rincón, Rodolfo G., *Contra una injusticia*, Caracas, Editorial Bolívar, 1936

Roldán Oliarte, Esteban, *El General Juan Vicente Gómez. Venezuela de cerca*, México, Imprenta Mundial, 1933

Torres, Gumersindo, *Memorándum de algunas Compañías petroleras sobre el Reglamento de la Ley sobre Hidrocarburos y demás minerales Combustibles y Observaciones que el Ministro de Fomento hace a dicho Memorándum*, Caracas, Tip. Central, 1930

Urbina, Rafael Simón, *Victoria, dolor y tragedia*, Caracas, Tip. Americana, 1936

Veloz Goiticoa, N., *Venezuela*, Caracas, Tip. Central, 1919

Articles

Adriani, Alberto, 'La crisis, los cambios y nosotros', *Cultura Venezolana*, 14:112 (Mayo–Junio 1931), 87–112

'La agricultura y la moneda', *BCCC*, 23:248 (Julio 1934), 6193–5

'Alemanes é Ingleses y el Petróleo Venezolano', *Boletín del Archivo Histórico de Miraflores*, 12:68 (Enero–Junio 1971), 131–75

'Apuntes sobre la situación cafetera venezolana', *Boletín del Banco Central de Venezuela*, 16:24 (Feb. 1947), 8–10

Biaggini, Angel, 'En torno de la Prima del Café', *Revista del Instituto Nacional del Café*, 1:1 (Agosto 1939), 63–70

Butler, J.D., 'The influences of economic factors on the location of oil refineries with primary reference to the world outside the USA and USSR', *The Journal of Industrial Economics*, 1:3 (July 1959), 187–201

Calvani, Luis F., 'Informe del Inspector Técnico de Minas al Ministerio de Fomento', *BCCC* 14:142 (1 Sept. 1925), 2979–80

'Circular dirigida a los Representantes de The Caribbean Petroleum Company, Lago Petroleum Corporation and West India Oil Company', *Boletín del Ministerio de Fomento* (Tercera Epoca), 1:6 (22 Enero 1931), 548–9

'Circular dirigida por el Ministro de Fomento a las Compañías Explotadoras de Petróleo, acerca de algunas dificultades encontradas para la inspección y fiscalización oficial', *Boletín del Ministerio de Fomento* (Tercera Epoca), 1:11 (1 Julio 1931), 1084–5

Córdoba, Armando, 'La estructura económica tradicional y el impacto petrolero en Venezuela', *Economía y Ciencias Sociales* (Caracas), 5:1 (Enero–Marzo 1963), 7–28

Crab, E.D., 'La Oficina Comercial de los Estados Unidos en Caracas', *BCCC*, 17:170 (1 Enero 1928), 3951

'Cuadros que manifestan la disminución de las deudas de Venezuela en los diez años comprendidos de 1 de enero de 1909 a 31 de diciembre de 1918', *BCCC*, 8:69 (1 Agosto 1919), 596–600

'Diógenes Escalante propone la participación venezolana en la explotación petrolera, 1926', *BAHM*, 13:70 (Enero–Feb. 1972), 347–52

Dupouy, Walter, 'El petróleo y las tierras agro-pecuarias', *El Farol*, 10:111 (Agosto 1948), 2–9

'Esteban Gil Borges, Canciller (1919–1920)', *BAHM* 15:76 (Julio–Dic. 1973), 147–226 (pp. 174–5)

González R., and Baldó, Lucio, 'Memorándum sobre la Compañía Venezolana de Petróleo, desde su fundación en junio de 1923 hasta el mes de octubre de 1920 *(sic)*', Caracas, 15.6.46. in Rafael González Rincones, *Cartas Barinesas*, Caracas, Editorial Sucre, 1958, 212–14

'Guillermo Zuloaga 1930–1970. Las Inspectorías Técnica de Hidrocarburos', *El Farol*, Suplemento al No. 234, ND

Henao Jaramillo, Jaime, 'La industria cafetera de Venezuela', *Revista del Instituto Nacional del Café*, 3:10 (Dic. 1941), 35–9

Lee, Thomas F., 'The rush for oil in Venezuela', *The World Today* (December 1925), 355–68

León, Ramón, 'Consideraciones sobre el café', *Revista del Instituto Nacional del Café*, 1:1 (Agosto 1939), 71–6

Linares, Oscar, 'Apreciaciones sobre la producción de café en Venezuela comparada con la de Colombia', *BCCC*, 14:143 (1 Oct. 1925), 3016–18

'Los Memorables de Gumersindo Torres', *BAHM*, 2:9 (Nov.–Dic. 1960), 157–65

Mieres, F., 'Los efectos de la explotación petrolera sobre la agricultura de

Venezuela', in Malavé Mata, Héctor, *Petróleo y Desarrollo económico de Venezuela*, La Habana, Publicaciones Económicas, 1964, 343–71

Morris, Henry C., 'Fomento de la Industria petrolera en las Américas', *Boletín del Ministerio de Fomento* (Segunda Epoca), 2:21 (Junio 1922), 943–7

'Pérez Soto y las Compañías petroleras, 1926', *BAHM*, 13:70 (Enero–Feb. 1972), 319–46

'Una poderosa industria nacional, la gasolina y el kerosere de Venezuela', *El Neuevo Diario* (Caracas, 15 Sept. 1917)

'Los Presidentes de Estado y los sucesos del año 28', *BAHM*, 2:7 (Julio–Agosto 1960) 123–4

'Progresos de la explotación de petróleo en el Zulia', *BCCC*, 14:136 (1 Marzo 1925), 2764

'Proyecto de Ley de Impuestos al Petróleo', *Boletín del Ministerio de Fomento*, 2:16 (Enero 1922), 721–8

'Proyecto de Ley – Ministerio de Fomento', *Boletín del Ministerio de Fomento*, 2:19 (Abril 1922), 862–75

'Proyectos e Intrigas petroleras (1926)', *BAHM*, 13:70 (Enero–Feb. 1972), 353–65

'Terrenos petrolíferos de Venezuela y Colombia', *BCCC*, 9:81 (15 Agosto 1920), 805–6

Urrutía, Claudio, 'Apuntes sobre la industria petrolera en Venezuela y su influencia en la vida económica del país', *BCCC*, 23:246 (Mayo 1934), 6147–50; and in *Boletín del Ministerio de Fomento* (Edición Extraordinario) (28 Julio 1934), 3039–49

'Venezuela en la Liga de las Naciones (1925)', *BAHM*, 13:70 (Enero–Feb. 1972), 207–39

'Venezuelan petroleum developments', *Petroleum Times* (London) 14:340 (11 July 1925), 51–2.

'El Zulia en 1926 y en 1929', *BAHM*, 17:90 (Marzo–Abril 1976), 53–77

Index

Abreu, Emilio, 139
Acosta, Henrique, 144
Adams, Paul, 165
Agriculture, 129–30, 133, 136; exports of, 129, 135; imports of, 135; products of, 110, 119, 138
Aguerrevere, E., 184
Aguerrevere, Jorge, 182
Alamo, Antonio, 59, 90–1, 104–6, 163, 165
Altagracia, 156
Alvarado, Lisandro, 7
Alvarado, Manuel, 93
Alvárez Cienfuegos, Felipe, 18, 30
Alvárez de Lugo, R., 183
Alvárez López Méndez, M.A., 30
Alves, Duncan Elliott, 12, 19, 27, 39–40
Amado Mejía, José, 33
Ambard, Félix R., 31
American-Venezuelan Development Corp., 95
American-Venezuelan Oilfields C.A., 102
Andes Petroleum Corp., 169
Andrade, Ignacio, 22, 31, 46, 67
Andrade, José, 46
Andújar, Antonio, 169–70
Anzola, Juvenal, 19
Apure-Venezuelan Petroleum Corp., 91, 97
Aranguren concession, 18, 41, 48, 161
Aranguren, Alberto, 139
Aranguren, Antonio, 12, 19, 27, 47, 102
Araujo, Rubén, 158
Arcaya, Pedro Manuel, 22, 28–30, 32–3, 36, 47, 52, 55, 59, 90–1, 96, 134, 136, 145, 147, 149–50, 154, 165–7, 176, 195, 204, 209
Arend Petro. My, 180
Aristimuño Coll, Carlos, 22
Armas, Sinforoso de, 76

Armstrong, R., 100
Arnold, Ralph, 12
Arroyo, Julián, 40
Arroyo Lameda, E., 197
Aspurúa Feo, R., 46
Astor Petroleum Co., 71
Atlantic Refining Co., 84
Ayala, Ramón, 31
Azuaje, Argenis, 151–2

Báez, Antonio, 46
Baldó, Lucio, 85, 98, 161, 168
Banco Agrícola y Pecuario, 121, 129–31
Banco de Venezuela, 89, 132, 134
Banco Mercantil y Agrícola, 96
Banco Urquijo, 169
Baptista Galindo, F., 83
Barco concession, 95
Barco Petroleum Corp., 48
Bariancas, 172
Barroso No. 2 well, 67
Bedford, A. C., 33
Bello Rodríguez, Dolores L. de, 83
Bennett, A. P., 98
Bermúdez Co., 10, 25
Betancourt, Simón, 151
Betancourt Sucre, J. M., 196
Bingham, R., 100, 165
Blanco, Carmen Lecuna de, 84
Blanco, Luis Felipe, 56
Booker, Leon C., 184
Bolivar Concessions Ltd, 12, 19
Bolivar Concessions (1917) Ltd, 19
Bolivar Exploration Co. Ltd, 83
Boscán, Acisclo, 150
Brandt, C. A., 46, 71
Braschi, J.M., 142, 160
Brice, Angel Francisco, 93
Briceño, M. S., 83
Britain, 109–10; oil companies of, 67

269

CAMBRIDGE LATIN AMERICAN STUDIES